Challenges and Techniques in Signal Detection with Unknown Covariance Matrix

Sana

Table of Contents

Appendices   101

Chapter 1: Introduction

1.1 Overview

The main focus of this **book** is on the detection of signals in the presence of additive Gaussian noise, in the case when the covariance matrix of the noise is unknown. While this problem itself is quite general, the signal and data models discussed in this **book** lend themselves readily to applications in adaptive radar systems, particularly in the detection of moving ground targets using airborne radar, which will be the focus of some brief expository notes in what follows.

1.2 Some Notes on Space-Time Adaptive Processing

1.2.1 Signal Model

In this section, we give a brief overview of the physical setup and the detection problem at hand. A more detailed description of STAP radar is provided in [91, 47], while the detection problem is covered in [56] and [80]. Assume we have a uniform linear array (ULA) with N antenna elements which receives M pulses across a datacube of N_b range bins. The spatial steering vector for a signal impinging on the ULA for one pulse at a normalized spatial frequency of ϑ is given as

1

$$\mathbf{a}_s(\vartheta) = \begin{bmatrix} 1 & e^{j2\pi\vartheta} & \dots & e^{j2\pi(N-1)\vartheta} \end{bmatrix}^T. \tag{1.1}$$

The corresponding temporal steering vector for the signal received for one element of the ULA across M pulses at a normalized temporal frequency of ω is given as

$$\mathbf{a}_t(\omega) = \begin{bmatrix} 1 & e^{j2\pi\omega} & \dots & e^{j2\pi(M-1)\omega} \end{bmatrix}^T. \tag{1.2}$$

Taken together, the received data matrix $\mathbf{X} \in \mathbb{C}^{N \times M}$ for one range cell of the STAP datacube consists of signal returns from N antennas and M pulses. The matrix $\mathbf{X}$ is typically used in its vectorized form which can be expressed as a linear combination of steering vectors of the form

$$\mathbf{a}(\omega, \vartheta) = \mathbf{a}_t(\omega) \otimes \mathbf{a}_s(\vartheta), \tag{1.3}$$

where $\otimes$ denotes the left Kronecker product, and of the complex white Gaussian noise vector $\mathbf{n} \sim \mathcal{CN}(0, \sigma^2 I)$.

1.2.2 Detection Problem

In classical STAP, a certain range cell is taken to be the Cell Under Test (CUT), and the detection problem is to decide whether or not the selected CUT contains a moving target. This process is repeated for all the range cells in the datacube. The detection problem can be stated as a choice between the hypotheses H_0 (target absent) and H_1 (target present), given as

$$H_0 \;:\; \mathbf{x} = \mathbf{c} + \mathbf{n} \tag{1.4}$$

$$H_1 \;:\; \mathbf{x} = b\mathbf{a}(\omega_t, \vartheta_t) + \mathbf{c} + \mathbf{n} \tag{1.5}$$

where the target has normalized temporal and spatial frequencies of ω_t, ϑ_t and an unknown complex amplitude b. The interference vector $\mathbf{x}_i = \mathbf{c} + \mathbf{n}$ is assumed to be distributed as $\mathbf{x}_i \sim \mathcal{CN}(0, \mathbf{R})$. A set of K closest range cells to the CUT, assumed to be target-free, statistically independent, and distributed identically as $\mathbf{x}_i$, are taken to be the training data. Of course, in practice ω_t, ϑ_t are not known ahead of time, so a commonly used approach is to form a group of candidate steering vectors parameterized by a discrete set of normalized temporal and spatial frequencies in the range $[-0.5, 0.5]$. A detection statistic incorporating the assumed set of steering vectors and the K training samples is then applied to the CUT, and the ω, ϑ values for which this statistic exceeds a certain threshold are declared to be the temporal and spatial frequencies at which a target is present.

One such statistic [56] compares the likelihood functions of the CUT and the training data for each of the hypotheses to form a generalized likelihood ratio test (GLRT) given as

$$\frac{|\mathbf{a}(\omega, \vartheta)^H \widehat{\mathbf{R}}^{-1} \mathbf{z}|^2}{\mathbf{a}(\omega, \vartheta)^H \widehat{\mathbf{R}}^{-1} \mathbf{a}(\omega, \vartheta)[1 + \frac{1}{K} \mathbf{z}^H \widehat{\mathbf{R}}^{-1} \mathbf{z}]} \mathop{\gtrless}_{H_0}^{H_1} \eta \qquad (1.6)$$

where $\widehat{\mathbf{R}} = \frac{1}{K} \sum_{k=1}^{K} \mathbf{z}_k \mathbf{z}_k^H$ is the sample covariance matrix formed from the training vectors $\mathbf{z}_1 \ldots \mathbf{z}_K \in \mathbb{C}^{NM}$, and $\mathbf{z}, \mathbf{a}(\omega, \vartheta)$ are the CUT and the steering vector for the normalized temporal and spatial frequencies ω and ϑ, respectively. Another similar statistic [6] forms the GLRT assuming the true clutter and noise covariance matrix is known. The true covariance is then replaced with $\widehat{\mathbf{R}}$ in an ad hoc manner, giving the Adaptive Matched Filter (AMF) statistic

$$\frac{|\mathbf{a}(\omega, \vartheta)^H \widehat{\mathbf{R}}^{-1} \mathbf{z}|^2}{\mathbf{a}(\omega, \vartheta)^H \widehat{\mathbf{R}}^{-1} \mathbf{a}(\omega, \vartheta)} \mathop{\gtrless}_{H_0}^{H_1} \eta. \qquad (1.7)$$

As the name implies, the squared magnitude of the numerator of this statistic is simply the output of CUT after being filtered with the vector $\mathbf{w} = \widehat{\mathbf{R}}^{-1}\mathbf{a}(\omega, \vartheta)$, the matched filter, which maximizes the signal to interference plus noise ratio

$$\text{SINR} = \frac{|\mathbf{w}^H \mathbf{a}(\omega, \vartheta)|^2}{\mathbf{w}^H \widehat{\mathbf{R}} \mathbf{w}}. \tag{1.8}$$

The denominator of the AMF statistic provides the useful Constant False Alarm Rate (CFAR) property, meaning that the false alarm rate given by this statistic is theoretically independent of the covariance matrix and of any scaling of the steering vectors or training data [56, 80]. It has been shown through numerical experiments that the AMF gives similar performance to the GLRT, and is equivalent to it in the asymptotic regime as the number of snapshots K approaches infinity.

1.2.3 Estimation Problem

In the derivation so far, it was assumed that the set of training vectors $\mathbf{z}_1 \ldots \mathbf{z}_K$ taken from nearby range cells that are target free and homogeneous. In real-world STAP applications, this is hard to come by for large K, due to factors like terrain variability, high concentration of targets, and others. Various approaches have been taken to combat this problem, such as outlier excision, modeling the clutter using heavy tailed general distributions, performing STAP directly on the CUT without training data, and many others. One approach which has been widely pursued in the literature is to simply assume that the target free and homogeneous property of the nearby range cells is maintained over a narrow range swath, corresponding to a small number of range cells next to the CUT. In practice, this means we need to estimate the clutter and noise covariance matrix using as few training vectors as possible. The main metric we use to judge the quality of our estimated covariance matrix is the normalized SINR (nSINR) given as

$$\text{nSINR} = \frac{|\mathbf{a}(\omega,\vartheta)^H\widehat{\mathbf{R}}^{-1}\mathbf{a}(\omega,\vartheta)|^2}{|\mathbf{a}(\omega,\vartheta)^H\widehat{\mathbf{R}}^{-1}\mathbf{R}\widehat{\mathbf{R}}^{-1}\mathbf{a}(\omega,\vartheta)||\mathbf{a}(\omega,\vartheta)^H\mathbf{R}^{-1}\mathbf{a}(\omega,\vartheta)|} \tag{1.9}$$

which is readily seen to be the ratio of the SINR using the estimated covariance matrix $\widehat{\mathbf{R}}$ to the optimal SINR with true covariance $\mathbf{R}$. Using the sample covariance matrix estimate for $\widehat{\mathbf{R}}$, it can be shown [78] that the nSINR follows a beta distribution parameterized by the matrix dimensionality NM and number of training vectors K, and that to get an expected nSINR loss that is half or less (i.e., within 3 dB) of the optimal, we need $K \geq 2NM$. Furthermore, the sample covariance matrix is singular if $K < NM$. For these reasons, numerous recent algorithms (e.g., [74, 36, 37, 85, 87, 79] and many others) have been proposed to estimate $\mathbf{R}$ in a manner superior to the standard sample covariance technique. This is usually done by incorporating various assumptions on its structure, some of which will be detailed in the next section.

1.3 Assumptions on the Structure of R

1.3.1 Basic Decomposition

It is often assumed (e.g., [91, 53, 79]) that the clutter and noise covariance matrix can be written as

$$\mathbf{R} = \mathbf{R}_c + \sigma_n^2\mathbf{I}$$

where σ_n^2 denotes the variance of the system thermal noise and $\mathbf{R}_c$ is a low rank matrix representing the clutter component. Many algorithms (e.g., [92, 53, 79], and many others) take σ_n^2 to be either known or have a lower bound on it; for radar systems this is usually a fairly practical assumption since the noise floor can be estimated by operating the radar in receive-only mode.

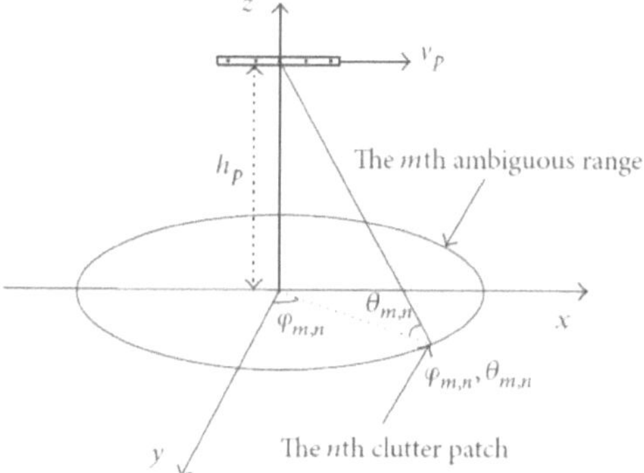

Figure 1.1: Airborne platform geometry [47].

1.3.2 Low Rankness of $\mathbf{R}_c$

To explain why exactly $\mathbf{R}_c$ turns out to be low rank, it is helpful to examine the geometry of the STAP problem in a little more detail. The basic setup is shown in Figure 1.1, where a ULA is mounted on an aircraft in a side-looking configuration. The normalized spatial frequency ϑ can now be expressed as

$$\vartheta = \frac{d}{\lambda_c}\sin(\phi)\cos(\theta) \tag{1.10}$$

where d is the spacing between antenna elements in the ULA and λ_c is the wavelength of operation. Likewise, the normalized temporal frequency ω can be expressed as

$$\omega = \frac{2v_p T_r}{\lambda_c}\sin(\phi)\cos(\theta) \tag{1.11}$$

where v_p is the platform velocity and T_r is the pulse repetition interval. It can be seen that the normalized spatial and temporal frequencies are related as

$$\omega = \beta\vartheta \tag{1.12}$$

6

where $\beta = \frac{2v_p T_r}{d}$. The clutter component $\mathbf{c}$ of the received signal can be approximately expressed [91] as the summation of clutter returns from a large number N_c of discrete clutter patches as

$$\mathbf{c} = \sum_{i=1}^{N_c} \sigma_i \mathbf{a}(\omega_i, \vartheta_i) \tag{1.13}$$

where σ_i is the random amplitude of the ith clutter patch. Assuming returns from the patches are uncorrelated, the clutter component of the covariance matrix can be written as

$$\mathbf{R_c} = \mathbb{E}[\mathbf{cc}^H] = \mathbf{V}_c \boldsymbol{\Sigma}_c \mathbf{V}_c^H \tag{1.14}$$

where $\boldsymbol{\Sigma}_c = \mathrm{diag}(\mathbb{E}[|\sigma_1|^2] \ldots \mathbb{E}[|\sigma_{N_c}|^2])$ and $\mathbf{V}_c = [\mathbf{a}(\omega_1, \vartheta_1) \ldots \mathbf{a}(\omega_{N_c}, \vartheta_{N_c})]$. Here, the linear relationship $\omega = \beta \vartheta$ between the normalized spatial and temporal frequencies can be exploited to reveal some interesting features about the matrix $\mathbf{V}_c$. When β is an integer, the steering vector $\mathbf{a}(\omega_i, \vartheta_i) = \mathbf{a}(\beta \vartheta_i, \vartheta_i)$ consists of multiples of the variable $z_i = \exp(j2\pi\vartheta_i)$, and the matrix $\mathbf{V}_c$ can be expressed as [91]

$$\mathbf{V}_c = \left[\begin{array}{cccc}
1 & 1 & \cdots & 1 \\
z_1 & z_2 & \cdots & z_{N_c} \\
\vdots & \vdots & & \\
z_1^{N-1} & z_2^{N-1} & \cdots & z_{N_c}^{N-1} \\
\hline
z_1^{\beta} & z_2^{\beta} & \cdots & z_{N_c}^{\beta} \\
z_1^{\beta+1} & z_2^{\beta+1} & \cdots & z_{N_c}^{\beta+1} \\
\vdots & \vdots & & \\
z_1^{\beta+N-1} & z_2^{\beta+N-1} & \cdots & z_{N_c}^{\beta+N-1} \\
\hline
\vdots & \vdots & & \\
\hline
z_1^{(M-1)\beta} & z_2^{(M-1)\beta} & \cdots & z_{N_c}^{(M-1)\beta} \\
z_1^{(M-1)\beta+1} & z_2^{(M-1)\beta+1} & \cdots & z_{N_c}^{(M-1)\beta+1} \\
\vdots & \vdots & & \\
z_1^{(M-1)\beta+N-1} & z_2^{(M-1)\beta+N-1} & \cdots & z_{N_c}^{(M-1)\beta+N-1}
\end{array} \right] \tag{1.15}$$

This structure follows from the Kronecker product form of the spatio-temporal steering vector, and the aforementioned relationship $\omega = \beta\vartheta$. The unique rows of $\mathbf{V}_c$ can now be grouped together as

$$\mathbf{V}_c = \begin{bmatrix} 1 & \cdots & 1 \\ z_1 & \cdots & z_{N_c} \\ z_1^2 & \cdots & z_{N_c}^2 \\ & \vdots\ \vdots & \\ z_1^{(M-1)\beta+N-1} & \cdots & z_{N_c}^{(M-1)\beta+N-1} \\ \hline & \vdots\ \vdots & \end{bmatrix} \tag{1.16}$$

The matrix $\mathbf{V}_c$ can now be seen to only have $N_r = N + \beta(M-1)$ unique rows. Because N_c is often assumed to be large, it can be seen that the rank of $\mathbf{V}_c$ (and by extension $\mathbf{R_c}$, because $\boldsymbol{\Sigma_c}$ is generally full rank) is given by $\min(N_r, N_c) = N_r$. This result is popularly referred to as Brennan's rule. [1]

In addition to the mathematical interpretation above, a physical interpretation of this rule can be seen from Figure 1.2. In this illustration, there are four antennas which capture signal snapshots across three pulses. Although there are twelve total observations, only six of them, the number predicted by Brennan's rule, are independent. The rest of them are approximately equal because the same signal is merely coming from different antennas at closely-spaced time intervals.

1.3.3 Impact of Crabbing Angle

The previous discussion focused on the case where the antenna array was aligned with the direction of motion of the aircraft platform. In general, this may not be the case; due to effects like crosswind, the aircraft may actually be traveling in a slanted manner such that its body and the attached antenna array are no longer aligned

---

[1]This is not to be confused with the Reed-Mallett-Brennan (RMB) rule [78], which relates to the amount of training data needed for adequate adaptive filter performance.

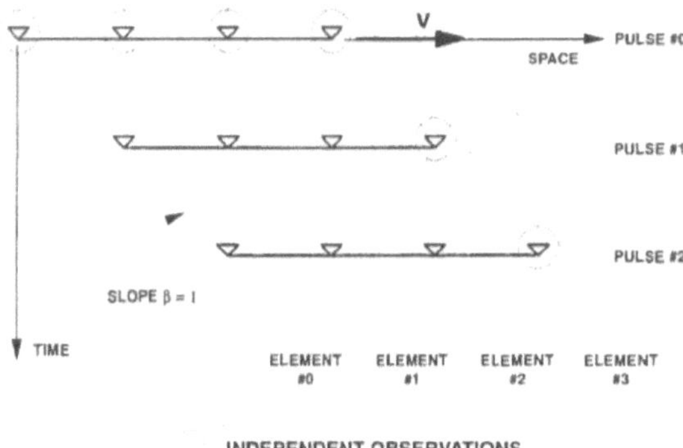

Figure 1.2: Illustration of Brennan's rule [91].

with the direction of motion. In STAP, this phenomenon is commonly referred to as crabbing angle. The presence of crabbing angle alters the relationship between the normalized spatial and temporal frequencies, which in turn affects the low rankness assumption on $\mathbf{R}_c$.

The scenario described is depicted in Figure 1.3. For a perfectly aligned antenna array with $\phi_a = 0$, the clutter patch at angle ϕ from boresight appears to be coming from the same angle as the clutter patch at the angle $180-\phi$, because the returns from both patches induce an identical phase shift on the array, from which the direction of arrival is determined. This is no longer the case when ϕ_a is nonzero. Moreover, due to the coupling between angle and Doppler in STAP, the Doppler returns from the two clutter patches in this case are also different. In this scenario, Brennan's rule no longer holds; however, in practice the effective rank of the clutter covariance

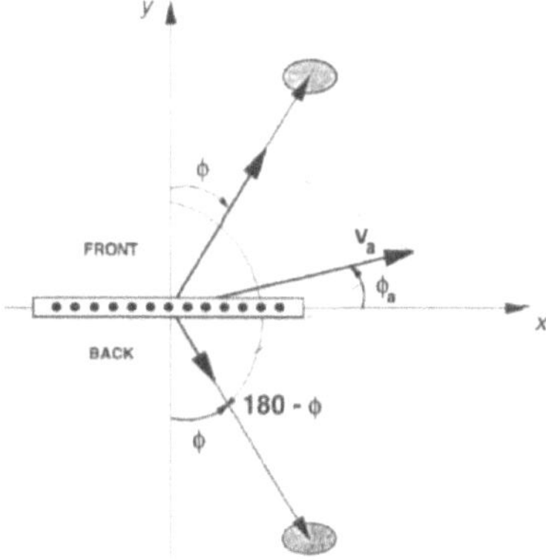

Figure 1.3: Illustration of a side-mounted ULA at crabbing angle ϕ_a [91].

matrix is approximately doubled from what Brennan's rule would predict. On the other hand, it must also be noted that the strength of the clutter returns may be significantly different because one corresponds to the frontlobe of the antenna array and the other to the backlobe. In theory, for an antenna array with a extremely low backlobe pattern, the clutter returns from the patch at the bottom of Figure 1.3 would be small and the effective rank predicted by Brennan's rule would not be changed as much, but in practice the backlobe component still needs to be accounted for [91].

The presence of antenna crabbing also alters the relationship between the normalized spatial and temporal frequencies. Without crabbing, the relationship between these two values is linear, as given by (1.11), meaning that the clutter ridge appears as a flat line in the spatio-temporal domain. However, when there is crabbing present, the relationship becomes more complicated, making the clutter ridge appear as an

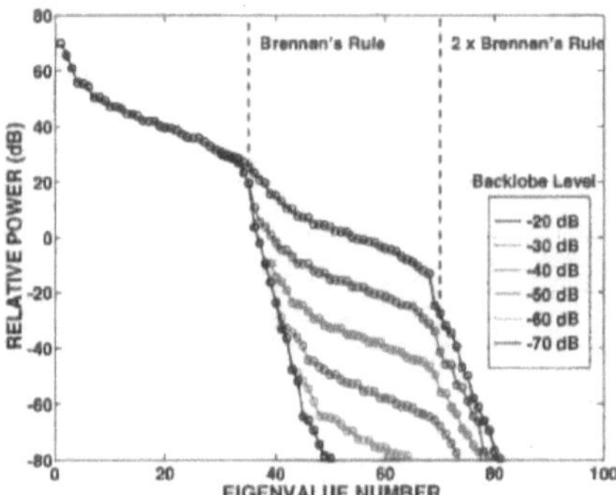

Figure 1.4: Impact of backlobe level on clutter covariance rank in presence of antenna crabbing [91].

ellipse. This is phenomenon is illustrated in Figure 1.4, and the new relationship between the normalized temporal and spatial frequencies can be approximately written as [91, 47]

$$\omega = \beta\vartheta\cos(\phi_a) \pm \beta\frac{d}{\lambda_c}\sin(\phi_a)\sqrt{1 - \sin^2(\sin^{-1}(\frac{\lambda_c}{d}\vartheta))} \tag{1.17}$$

where the "+" and "−" components in the equation correspond to the frontlobe and backlobe components of the antenna array returns, respectively.

1.3.4 Other Nonideal Effects: Internal Clutter Motion and Antenna Calibration Errors

Using the notation developed so far, the received target-free clutter and noise vector at one range cell can be expressed as

$$\mathbf{x} = \mathbf{c} + \mathbf{n} = \mathbf{V}_c\sigma + \mathbf{n} \tag{1.18}$$

11

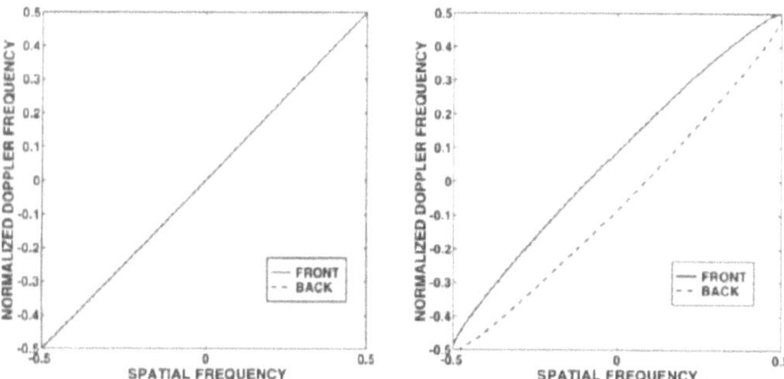

Figure 1.5: Clutter ridge in the angle-Doppler (normalized spatial and temporal frequency) domain, with (right) and without (left) antenna crabbing [91].

where $\sigma = [\sigma_1 \ \ldots \ \sigma_{N_c}]^T$. However, in most practical applications, there are other error sources which must be accounted for. One of these is internal (or intrinsic) clutter motion (ICM), which causes a fluctuation of the received signal from pulse to pulse. ICM, as its name suggests, is often caused by motion of the clutter that is independent from the relative motion induced by the moving platform, such as the swaying of trees and grass in the wind and the motion of waves and ocean currents in maritime applications. Another effect is antenna calibration errors, which for the time being are modeled as being angle independent, because the main beam is assumed to be fixed for the duration of data collection in the datacube (the coherent processing interval, or CPI). When incorporating these effects into the signal model, the received clutter and noise vector can now be written as [92]

$$\mathbf{x} = \mathbf{c} + \mathbf{n} = [(\mathbf{V_c} \odot \Xi_\mathbf{s})\sigma] \odot [\alpha_t \otimes \mathbf{1}_N] + \mathbf{n} \tag{1.19}$$

where $\boldsymbol{\Xi_s}$ is a matrix whose columns are all equal to $\mathbf{1}_M \otimes \alpha_s$. Here, α_t and α_s are the tapering vectors representing the temporal signal fluctuation across pulses and spatial fluctuation across antenna elements, respectively (note that the authors in [3], different from here, used the somewhat less common notation of N for number of pulses and M for number of antennas). It is assumed that α_t is random, but that α_s is constant across the CPI. The total clutter and noise covariance matrix can then be written as

$$\mathbf{R} = \mathbf{R}_s \odot \mathbf{T}_t + \sigma_n^2 \mathbf{I} \tag{1.20}$$

where $\mathbf{R}_s = (\mathbf{V_c} \odot \boldsymbol{\Xi_s})\boldsymbol{\Sigma_c}(\mathbf{V_c} \odot \boldsymbol{\Xi_s})^H$ and $\mathbf{T}_t = \mathbb{E}[\alpha_t\alpha_t^H] \otimes \mathbf{1}_N\mathbf{1}_N^H$. One of the main conclusions from this result is that when practical system imperfections are taken into account, the clutter component of the covariance matrix can no longer be expressed as a simple sum of steering vector outer products. This poses a considerable challenge for algorithms which solely depend on the outer product form of the clutter covariance matrix. However, if the error models for the ICM and antenna calibration errors remain constant across the range cells used for the training data, these imperfections are already accounted for in the sample covariance matrix. The issue of ICM, along with numerous other real-world impairments, illustrates some of the many challenges and considerations which must be taken into account in practical radar signal processing systems.

1.4 Organization of this book

Much of the background material and literature review for the problems discussed in this book is given within the relevant chapters themselves. The rest of this book is organized as follows.

Chapter 2 discusses covariance matrix estimation under various constraints and/or under a Bayesian prior. The main contribution of this chapter is the extension of a simple, closed-form covariance estimator which incorporates the physically-motivated constraints of low rank and known noise floor, as well as other constraints like per-symmetry and/or a Bayesian prior. This chapter is largely based on the results in [83].

Chapter 3 discusses adaptive detection in the presence of partial homogeneity when the condition number of the true covariance matrix is low. The nature of the recently introduced Adaptive Orthogonal Rejection Detector (AORD), to which the proposed detectors are related, is elucidated, and it is shown that both the AORD and the proposed detectors are white noise matched filters with a proper scaling to ensure Constant False Alarm Rate (CFAR). A performance trade-off is discussed which shows that in low training data scenarios, the use of the proposed detectors gives significantly improved performance over traditional detectors which use pseudowhitened matched filtering.

Chapter 4 proposes several detectors, designed under the framework of Generalized Likelihood Ratio (GLR), Rao, and Wald tests, which incorporate the assumption of spectral symmetry on the received colored Gaussian noise. The detectors are derived for both homogeneous and partially homogeneous scenarios and extend previously derived detectors from the point-like, single-vector case to the case of range-spread and/or subspace signals. In addition, detectors using the approximate one-step GLR and Wald tests are given. Numerical simulations are provided to investigate the performance of the proposed detectors in various regimes.

Chapter 5 discusses adaptive detection in the scenario when no secondary data is present, but the additive noise in the primary data is assumed to be spectrally symmetric. Detection strategies are presented on the basis of Canonical Correlation Analysis (CCA), and the presented detectors are shown to be robust with respect to changes in the underlying scaling factors (for the non-Gaussian noise case) and the covariance (for both Gaussian and non-Gaussian cases).

Chapter 2: Low-Rank Structured Covariance Matrix Estimation

2.1 Overview

In this chapter, covariance matrix estimates are presented in both the frequentist and Bayesian settings as the solutions of a maximum likelihood (ML) or maximum a posteriori (MAP) optimization, respectively, when the true covariance consists of a known (or bounded) noise floor and low-rank component. Persymmetric structure may also be assumed. The ML and MAP solutions with the non-convex rank constraint are shown to be a simple scalar thresholding of eigenvalues of a suitably translated and projected sample covariance matrix. No iterative optimization is required; therefore, the computation is suited to real-time applications. Our proof is short and elementary without resorting to duality theory. Numerical results are presented to illustrate the improved estimation performance obtained by incorporating the structural constraints on the unknown covariance matrix.

2.2 Introduction

The detection of a signal of interest among correlated interference is a problem of paramount importance in airborne radar systems [47], [91]. The canonical approach for the detection of such a signal is the one-step [56] or two-step [80] generalized likelihood ratio test (GLRT), which can similarly be thought of as the matched filter in colored Gaussian noise. However, this approach requires the knowledge of the covariance matrix of the interference signals, which consist of unwanted clutter returns, electronic jammers, and thermal noise. This covariance matrix is generally unknown and must be estimated from the nearby range cells in the radar datacube, assumed to be statistically homogeneous and target-free; however, in practice the secondary data are not ideal and are often contaminated with clutter heterogeneities and/or targets, resulting in few homogeneous training samples being available for use in covariance matrix estimation.

Various approaches have been proposed to deal with this challenge, often by exploiting available knowledge about the covariance matrix in the adaptive detection scenario. For example, when a uniform linear array (ULA) is used and the clutter returns come from uncorrelated sources, the resulting clutter and noise covariance matrix is Toeplitz structured; the same result occurs when the detection problem is viewed from the perspective of a single antenna using uniformly spaced time samples in pulse-Doppler processing. In the space-time adaptive processing (STAP) scenario with multiple pulses and antennas, and in the absence of systems-level imperfections (array calibration errors, etc.), the resulting covariance matrix is Toeplitz-block-Toeplitz (TBT). Some of the prior works which invoke the Toeplitz constraint include [37], [54], and [9], while those invoking persymmetry include [74], [30], and [43]; furthermore, Toeplitz and TBT matrices may be embedded into circulant matrices [36]. In this work, in order to obtain closed-form solutions to the MAP and ML problems without resorting to computationally intensive iterative approaches, an approximation to the Toeplitz and TBT structures is used in the form of the persymmetric structure.

Another approach that has been considered for improving covariance matrix estimation in the low-snapshot regime is the Bayesian approach, most commonly invoking a complex inverse Wishart prior on the covariance matrix [29, 28, 90]. The physical meaning behind the use of this approach in STAP is that in many real-world scenarios, there is available knowledge of the parameters relevant to predicting the structure of the covariance matrix, such as platform velocity, crabbing angle, digital terrain data, etc., which can be used to construct a prior matrix $\mathbf{R_0}$ that characterizes the inverse Wishart distribution. For the Wishart prior, the resulting MAP estimate is simply a scaled sum of the unbiased sample covariance matrix and the prior matrix $\mathbf{R_0}$. This result can be interpreted as a "colored loading," namely, the linear combination of the sample covariance matrix with a non-diagonal matrix [40, 87, 79]. Likewise, colored loading itself can be viewed as a subset of "shrinkage," which is the linear combination of the sample covariance matrix with some positive definite matrix [61, 85, 18, 79]. In a spirit similar to MAP estimation, convex programming may be used to determine a constrained covariance matrix closest to a specified positive semidefinite matrix [27, 4, 3]; adopting the spectral norm for distance, a lower bound on white thermal noise power, and an upper bound on condition number results in simple computation [3], rather than a computationally expensive iterative solution to the convex program.

A further approach has been to incorporate low-rankness, a non-convex constraint, into the covariance estimation problem. Due to the coupling between the azimuthal angle and the Doppler shift in airborne radar, the covariance of the clutter returns lies in a low-dimensional subspace whose rank can be predicted from system parameters (e.g., number of pulses and antennas, pulse repetition frequency, etc.) via Brennan's rule [91]. Thus, the interference covariance matrix can be decomposed into a low-rank clutter component and a scaled identity matrix representing the thermal noise, which is present in all electronic devices. Furthermore, the variance of the thermal noise itself can be estimated by running the radar system on receive-only mode when no active electronic interference is present.

Existing MAP formulations [29, 28, 90] admit a closed-form solution, but do not constrain the resulting estimate to conform to the physical constraints of low-rank clutter or an interval bound on thermal noise power. Likewise, closed-form solutions have been widely reported for maximum likelihood estimation for known thermal noise [84, 40, 5], or for both known thermal noise and low-rank clutter [2, 15, 53]; however, to the best of the authors' knowledge, no closed-form estimator has been reported when an additional symmetry constraint is enforced on the ML solution.

In this letter, we consider both MAP and ML estimation scenarios with simple, exact computation suitable for real-time implementation and yielding a performance improvement in many realistic scenarios relevant to airborne radar. We consider the structured covariance model, $\mathbf{R} = \mathbf{M} + \sigma^2 \mathbf{I}$, with $\mathbf{M}$ low rank and an interval bound on σ^2. Estimators are also provided which additionally enforce persymmetry on $\mathbf{R}$. Our analysis is elementary and self-contained, and does not require duality theory, which might be of independent interest.

Notation: For matrices, we use boldface uppercase letters, and we use $\mathrm{tr}(\mathbf{A})$ and $|\mathbf{A}|$ to denote the trace and determinant. We use superscripts ^{T}, ^{H}, and * for transpose, conjugate transpose, and conjugation, respectively; $\|\cdot\|_F$ and $\|\cdot\|_2$ are the Frobenius and spectral norms. For vectors, we use boldface lower case letters; scalars are in non-bold font. We use $\mathrm{diag}(\mathbf{a})$ to denote a square diagonal matrix with entries given by $\mathbf{a}$, and $\mathrm{diag}(\mathbf{A})$ to denote the diagonal entries taken from a square matrix $\mathbf{A}$; $[\mathbf{A}]_{kl}$ is the (k, l)th entry of $\mathbf{A}$, while a_k is the kth entry of a column vector, $\mathbf{a}$. Let $\mathbf{J}$ denote the p-by-p exchange matrix with entries $[\mathbf{J}]_{ij} = 1$ for $j = p - i + 1$ and zero otherwise, and let $\mathbf{I}$ denote the identity matrix, with the sizes of both matrices implied by context. Note that $\mathbf{J} = \mathbf{J}^T = \mathbf{J}^{-1}$. A matrix $\mathbf{M}$ is said to be persymmetric if $\mathbf{MJ} = \mathbf{JM}^T$. Finally, let $A \succeq B$ denote that $A - B$ is a positive semi-definite (psd) matrix.

2.3 Preliminaries

Consider iid, multivariate, complex circular Gaussian samples, $\mathbf{x}_k \sim \mathcal{CN}(\mathbf{0}, \mathbf{R})$ with zero mean and covariance matrix $\mathbf{R}$. Collect n samples (or, "snapshots") as columns in a data matrix, $\mathbf{X} \in \mathbb{C}^{p \times n}$. The joint density for $\mathbf{X}$ is

$$g(\mathbf{X}; \mathbf{R}) = \pi^{-pn} |\mathbf{R}^{-1}|^n \exp\left\{-\mathrm{tr}\left(\mathbf{R}^{-1}\mathbf{X}\mathbf{X}^H\right)\right\}. \tag{2.1}$$

The negative of the logarithm yields the ML risk

$$\mathrm{tr}\left\{\mathbf{R}^{-1}\mathbf{X}\mathbf{X}^H\right\} - n\log|\mathbf{R}^{-1}| \tag{2.2}$$

where constants unrelated to $\mathbf{X}$ and $\mathbf{R}$ have been dropped. This risk function is convex in the precision matrix, $\mathbf{R}^{-1}$.

In contrast, a MAP estimator can be defined given a prior on the covariance, $\mathbf{R}$. To this end, we adopt a complex central inverse Wishart prior, which is a conjugate prior density for the covariance matrix in a multivariate normal model. For l samples, $\mathbf{z}_k \in$

$\mathbb{C}^p$, drawn iid from $\mathcal{CN}(\mathbf{0}, \mathbf{\Sigma})$, the matrix $\mathbf{W} = \sum_{k=1}^{l} \mathbf{z}_k \mathbf{z}_k^H$ is complex central Wishart, $\mathcal{CW}(\mathbf{\Sigma}, l)$ [45, 82]. The matrix $\mathbf{A} = \mathbf{W}^{-1}$ has inverse Wishart distribution [86], $\mathcal{CIW}(\mathbf{\Sigma}^{-1}, l)$, with probability density proportional to $|\mathbf{A}|^{-(l+p)} \exp\{-\operatorname{tr}(\mathbf{\Sigma}^{-1}\mathbf{A}^{-1})\}$. To incorporate the low-rank structure and an interval bound on the thermal noise power, we define a set $\mathcal{S} = \{\mathbf{R} : \mathbf{R} = \mathbf{M} + \sigma^2 \mathbf{I}, \ \mathbf{M} \succeq 0, \ \operatorname{rank}(\mathbf{M}) \leq r\}$, with $\mathrm{LB} \leq \sigma^2 \leq \mathrm{UB}$ and $1 \leq r \leq p$; additionally, let $\mathcal{T}$ denote the set of persymmetric p-by-p matrices. We define a modified $\mathcal{CIW}(\mathbf{R}_0, l)$ distribution for the covariance matrix that is only nonzero on the set $\mathcal{S} \cap \mathcal{T}$:

$$f(\mathbf{R}; \theta) \begin{cases} \propto |\mathbf{R}|^{-(l+p)} \exp\{-\operatorname{tr}(\mathbf{R}_0 \mathbf{R}^{-1})\} & \mathbf{R} \in \mathcal{S} \cap \mathcal{T} \\ = 0, & \mathbf{R} \notin \mathcal{S} \cap \mathcal{T}. \end{cases} \tag{2.3}$$

where $\theta = \{\mathbf{R}_0, l, r, \mathrm{LB}, \mathrm{UB}\}$. The MAP estimate of $\mathbf{R}$ is then given as the solution of the constrained optimization problem

$$\arg\max_{\mathbf{R} \in \mathcal{S} \cap \mathcal{T}} \ g(\mathbf{X}; \mathbf{R}) f(\mathbf{R}; \theta). \tag{2.4}$$

The negative of the logarithm yields the MAP risk

$$\operatorname{tr}\left\{\mathbf{R}^{-1}(\mathbf{X}\mathbf{X}^H + \mathbf{R}_0)\right\} - (n + l + p) \log|\mathbf{R}^{-1}|. \tag{2.5}$$

The following lemma is an extension of the von Neumann trace inequality.

Lemma 1. *[72, 88] Let $\mathbf{A}$ and $\mathbf{B}$ be psd Hermitian matrices, with $\mathbf{A} = \mathbf{\Phi} \operatorname{diag}(\boldsymbol{\lambda_A})\mathbf{\Phi}^H$ and $\mathbf{B} = \mathbf{\Psi} \operatorname{diag}(\boldsymbol{\lambda_B})\mathbf{\Psi}^H$, where $\boldsymbol{\lambda_A}$ $(\boldsymbol{\lambda_B})$ are the eigenvalues, in non-increasing order, of $\mathbf{A}$ $(\mathbf{B})$, and $\mathbf{\Phi}$, $\mathbf{\Psi}$ are unitary. Then,*

$$\boldsymbol{\lambda_A}^T \mathbf{J} \boldsymbol{\lambda_B} \leq \operatorname{tr}(\mathbf{AB}) \leq \boldsymbol{\lambda_A}^T \boldsymbol{\lambda_B} \tag{2.6}$$

with equality at the lower (upper) bound if and only if $\mathbf{\Psi} = \mathbf{\Phi}\mathbf{J}$ $(\mathbf{\Psi} = \mathbf{\Phi})$.

The trace inequality yields the following result for optimizing a cost function over the closed, but non-convex, set of p-by-p psd matrices with rank not exceeding $r \leq p$ and with thermal noise power constrained to the interval $\text{LB} \leq \sigma^2 \leq \text{UB}$.

Lemma 2. *Let $\alpha > 0$, $\mathbf{S} = \mathbf{U}\mathbf{D}\mathbf{U}^H$ be psd, $\mathbf{D} = \text{diag}\,(d_1, \cdots, d_p)$, $d_1 \geq d_2 \geq \cdots \geq d_p \geq 0$, and r an integer, $1 \leq r \leq p$. Then, the solution to the optimization problem*

$$\underset{\mathbf{M} \succeq 0, \, \sigma^2}{\arg\min} \, \text{tr}\left\{(\mathbf{M} + \sigma^2\mathbf{I})^{-1}\mathbf{S}\right\} - \alpha \log\left|(\mathbf{M} + \sigma^2\mathbf{I})^{-1}\right| \tag{2.7}$$

$$s.t. \quad \text{rank}(\mathbf{M}) \leq r, \quad and \quad LB \leq \sigma^2 \leq UB$$

yields $(\widehat{\mathbf{M}} + \widehat{\sigma}^2\mathbf{I}) = \mathbf{U}\mathbf{\Lambda}\mathbf{U}^H$, where $\mathbf{\Lambda} = \text{diag}\,(\boldsymbol{\lambda})$ and

$$\lambda_k = \begin{cases} \max\left(\frac{d_k}{\alpha}, \ LB\right) & k = 1, \cdots, r \\ \min\left(\max\left(\frac{\bar{d}}{\alpha}, \ LB\right), \ UB\right) & k = r+1, \cdots, p. \end{cases} \tag{2.8}$$

Here, $\bar{d}$ is the average of the noise space eigenvalues, $\bar{d} = \frac{1}{p-r}\sum_{k=r+1}^{p} d_k$.

Lemma 2 generalizes the results in [15, 53] to an interval bound on σ^2; for completeness and independent interest, we present a short and elementary proof here.

Proof. Define $\mathbf{P} = (\mathbf{M} + \sigma^2\mathbf{I})^{-1}$ and rewrite the objective function as $\text{tr}\{\mathbf{P}\mathbf{S}\} - \alpha \log|\mathbf{P}|$. Using eigendecompositions, let $\mathbf{S} = \mathbf{U}\mathbf{D}\mathbf{U}^H$ and $\mathbf{P} = \mathbf{V}\mathbf{\Lambda}^{-1}\mathbf{V}^H$, where $\mathbf{D} = \text{diag}(d_1, \ldots, d_p)$ with $d_1 \geq \cdots \geq d_p$, and $\mathbf{\Lambda} = \text{diag}(\lambda_1, \ldots, \lambda_p)$ with $\lambda_1 \geq \cdots \geq \lambda_p$. Utilizing the trace inequality in Lemma 1, we can write

$$\text{tr}\{\mathbf{P}\mathbf{S}\} - \alpha \log|\mathbf{P}| \geq \alpha \left(\sum_{k=1}^{p} \frac{d_k/\alpha}{\lambda_k} + \log \lambda_k\right), \tag{2.9}$$

where the equality holds if and only if $\mathbf{U} = \mathbf{V}$. Here, $\lambda_{r+1} = \lambda_{r+2} = \cdots = \lambda_p$ by hypothesis, and $\bar{d} = \frac{1}{p-r}\sum_{k=r+1}^{p} d_k$. Observe the sum of scalar functions, $h(x; c) = \frac{c}{x} + \log x$, inside the parentheses of the right-hand term, and note that $h(x; c)$ is increasing in c for $x > 0$. Thus, by the nonincreasing ordering of the d_k values and the iid thermal noise model, the optimization with rank constraint on $\mathbf{M}$ reduces to

$$\underset{\lambda_1,\cdots,\lambda_{r+1}}{\arg\min} \sum_{k=1}^{r} \left(\frac{d_k/\alpha}{\lambda_k} + \log \lambda_k\right) + \sum_{k=r+1}^{p} \left(\frac{d_k/\alpha}{\lambda_{r+1}} + \log \lambda_{r+1}\right)$$

$$= \underset{\lambda_1,\cdots,\lambda_{r+1}}{\arg\min} \sum_{k=1}^{r} h(\lambda_k; \frac{d_k}{\alpha}) + (p-r)h\left(\lambda_{r+1}; \frac{\overline{d}}{\alpha}\right) \tag{2.10}$$

subject to $\text{LB} \le \lambda_k$, $k = 1, \cdots, r+1$ and $\lambda_{r+1} \le \text{UB}$. Observe that $h(x; c)$ is strictly decreasing in x for $x < c$ and increasing for $x > c$. Thus, to minimize each $h(x; c)$ in Eq. (2.10) $\lambda_k = \max\left(\frac{d_k}{\alpha}, \text{LB}\right)$ for $k = 1, ..., r$ and $\lambda_{r+1} = \min\left(\max\left(\frac{\overline{d}}{\alpha}, \text{LB}\right), \text{UB}\right)$.

$\square$

Let $\phi(\mathbf{d}; \alpha, \text{LB}, \text{UB}, r)$, or the shorthand $\phi(\mathbf{d})$, denote the application of the thresholding in Eq. (2.8) to a vector $\mathbf{d}$.

As consequences of Lemma 2, we incorporate constraints into MAP and ML structured covariance estimation problems; the covariance is of the form $\mathbf{M} + \sigma^2\mathbf{I}$ with low-rank $\mathbf{M}$ or low-rank persymmetric $\mathbf{M}$. In each case, the optimization admits a simple, closed-form solution despite the non-convexity of the rank constraint.

2.4 Rank-constrained MAP

First consider constrained MAP estimators for a structured covariance matrix.

Proposition 1 (Rank-constrained MAP). *Given iid, zero mean, multivariate complex circular Gaussian observations* $\mathbf{X} \in \mathbb{C}^{p \times n}$, $LB \le \sigma^2 \le UB$, *an inverse Wishart prior* $\mathcal{CIW}(\mathbf{R}_0, l)$, $l \ge p$, *and rank* r, $1 \le r \le p$, *the maximum a posteriori probability covariance estimate,* $\widehat{\mathbf{R}} = \mathbf{M} + \sigma^2\mathbf{I}$, *with* $\text{rank}(\mathbf{M}) \le r$ *and* $\mathbf{M} \succeq 0$, *is* $\widehat{\mathbf{R}} = \mathbf{U}\,\text{diag}(\phi(\boldsymbol{\lambda}))\mathbf{U}^H$ *with* $\alpha = n + l + p$ *and eigendecomposition* $\left(\mathbf{X}\mathbf{X}^H + \mathbf{R}_0\right) = \mathbf{U}\,\text{diag}(\boldsymbol{\lambda})\mathbf{U}^H$.

Proof. The MAP risk from Eq. (2.5) is equivalent to the cost in Eq. (2.7), with $\alpha = n + l + p$ and $\mathbf{S} = \mathbf{X}\mathbf{X}^H + \mathbf{R}_0$. Thus, let $\mathbf{U}\operatorname{diag}(\mathbf{d})\mathbf{U}^H$ be the eigendecomposition of the scaled and translated sample covariance matrix, $\mathbf{X}\mathbf{X}^H + \mathbf{R}_0$. Then, by Eq. (2.8) in Lemma 2, the MAP estimate is $\mathbf{U}\operatorname{diag}(\phi(\mathbf{d}))\mathbf{U}^H$. $\qquad\square$

Thus, the MAP covariance estimate, $\widehat{\mathbf{R}}$, is obtained by applying the thresholding operator of Eq. (2.8) to the eigenvalues of the scaled sample covariance matrix, $\mathbf{X}\mathbf{X}^H$, translated by the inverse Wishart scale matrix, $\mathbf{R}_0$. Note that $\mathbf{X}\mathbf{X}^H$ is n times the sample covariance matrix, and thus grows large as n increases, thereby increasing reliance of the MAP estimate on the data samples, rather than on the prior.

Proposition 2 (Rank-constrained Persymmetric MAP). *Given iid, zero mean, multivariate complex circular Gaussian observations $\mathbf{X} \in \mathbb{C}^{p \times n}$, $LB \leq \sigma^2 \leq UB$, an inverse Wishart prior $\mathcal{CIW}(\mathbf{R}_0, l)$, $l \geq p$, and rank r, $1 \leq r \leq p$, the maximum a posteriori probability covariance estimate, $\widehat{\mathbf{R}} = \mathbf{M} + \sigma^2 \mathbf{I}$, with $\operatorname{rank}(\mathbf{M}) \leq r$ and $\mathbf{M} \succeq 0$ persymmetric, is $\widehat{\mathbf{R}} = \mathbf{U}\operatorname{diag}(\phi(\mathbf{d}))\mathbf{U}^H$ with $\alpha = n + l + p$ and eigendecomposition $\mathbf{S} = \mathbf{U}\operatorname{diag}(\mathbf{d})\mathbf{U}^H$ for*

$$\mathbf{S} = \tfrac{1}{2}\left\{ \left(\mathbf{X}\mathbf{X}^H + \mathbf{R}_0\right) + \mathbf{J}\left(\mathbf{X}\mathbf{X}^H + \mathbf{R}_0\right)^T \mathbf{J}\right\}. \tag{2.11}$$

Proof. An invertible persymmetric matrix has persymmetric inverse; so, let $\mathbf{P} = (\mathbf{M} + \sigma^2\mathbf{I})^{-1}$ be a persymmetric matrix. Then, defining $\mathbf{S}_0 = \mathbf{X}\mathbf{X}^H + \mathbf{R}_0$, we have

$$\operatorname{tr}\left\{\mathbf{P}\mathbf{S}_0\right\} = \frac{1}{2}\left\{\operatorname{tr}\left\{\mathbf{P}\mathbf{S}_0\right\} + \operatorname{tr}\left\{\mathbf{J}\mathbf{P}^T\mathbf{J}\mathbf{S}_0\right\}\right\}$$

$$= \tfrac{1}{2}\operatorname{tr}\left\{\mathbf{P}\left\{\mathbf{S}_0 + \mathbf{J}\mathbf{S}_0^T\mathbf{J}\right\}\right\} = \operatorname{tr}\left\{\mathbf{P}\mathbf{S}\right\}$$

where we utilized the invariance property of the trace operation to cyclic permutations and transposition. Let $\mathbf{U}\operatorname{diag}(\mathbf{d})\mathbf{U}^H$ be the eigendecomposition of $\mathbf{S}$ from Eq. (2.11);

by Lemma 2, the rank-constrained MAP risk is minimized by $\widehat{\mathbf{R}} = \mathbf{U}\operatorname{diag}\{\phi(\mathbf{d})\}\mathbf{U}^H$. It remains only to show that the Hermitian matrix $\widehat{\mathbf{R}}$ is persymmetric. To this end, for distinct eigenvalues let $\{\mathbf{u}, d\}$ be an eigenvector/eigenvalue pair for $\mathbf{S}$ and note

$$d\mathbf{u} = \mathbf{S}\mathbf{u} = \mathbf{J}\mathbf{S}^*\mathbf{J}\mathbf{u} \Rightarrow \mathbf{S}^*\mathbf{J}\mathbf{u} = d\mathbf{J}\mathbf{u},$$

implying that $\{\mathbf{J}\mathbf{u}, d^*\}$ is also an eigenvector/eigenvalue pair. Because d is real-valued, we learn that $\mathbf{u}$ is either symmetric ($\mathbf{J}\mathbf{u} = \mathbf{u}^*$) or skew-symmetric ($\mathbf{J}\mathbf{u} = -\mathbf{u}^*$). Thus, $\mathbf{u}\mathbf{u}^H = \mathbf{J}\mathbf{u}^*\mathbf{u}^T\mathbf{J} = \mathbf{J}(\mathbf{u}\mathbf{u}^H)^*\mathbf{J}$ is persymmetric. The same conclusion holds for the case of repeated eigenvalues [75]. Thus, because $\phi(d_i)$ is real-valued, it follows that $\widehat{\mathbf{R}} = \sum_{i=1}^{p} \phi(d_i)\mathbf{u}_i\mathbf{u}_i^H$ is the sum of rank-1 persymmetric matrices and hence persymmetric. $\qquad\square$

Note that $\frac{1}{2n}\left\{\mathbf{X}\mathbf{X}^H + \mathbf{J}\left(\mathbf{X}\mathbf{X}^H\right)^T\mathbf{J}\right\}$ is the projection of the sample covariance matrix onto the closed, convex set of Hermitian persymmetric matrices [75]. So, the low-rank persymmetric MAP covariance estimate, $\widehat{\mathbf{R}}$, is obtained by applying the thresholding operator of Eq. (2.8) to the eigenvalues of the scaled sample covariance matrix, $\mathbf{X}\mathbf{X}^H$, translated by the inverse Wishart scale matrix, $\mathbf{R}_0$, and projected to the set of persymmetric Hermitian matrices. Thus, fast computation is retained while leveraging prior knowledge of the covariance, its thermal noise floor, low-rank clutter, and covariance structure. Further, the persymmetry may be exploited for computational efficiency in computing STAP weight vectors using the estimated covariance, $\widehat{\mathbf{R}}$ [52, 32, 33].

Remark 1. *The result for persymmetric structure immediately generalizes to* **T**-*conjugate symmetry. Let* $\mathbf{T} \in \mathbb{C}^{p\times p}$ *be a nontrivial involution; i.e.,* $\mathbf{T} = \mathbf{T}^{-1} \neq \pm\mathbf{I}$. *A matrix* **R** *is said to be* **T**-*conjugate [89] if* $\mathbf{R} = \mathbf{T}\mathbf{R}^*\mathbf{T}$.

2.5 Rank-constrained ML

We next consider similar extension of ML estimators.

Proposition 3 (Rank-constrained ML). *Given iid, zero mean, multivariate complex circular Gaussian observations* $\mathbf{X} \in \mathbb{C}^{p \times n}$, $LB \leq \sigma^2 \leq UB$, *and rank* r, $1 \leq r \leq p$, *the maximum likelihood covariance estimate,* $\widehat{\mathbf{R}} = \mathbf{M} + \sigma^2 \mathbf{I}$, *with* $\mathrm{rank}(\mathbf{M}) \leq r$ *and* $\mathbf{M} \succeq 0$, *is* $\widehat{\mathbf{R}} = \mathbf{U}\,\mathrm{diag}(\phi(\mathbf{d}))\mathbf{U}^H$ *with* $\alpha = n$ *and eigendecomposition* $\mathbf{X}\mathbf{X}^H = \mathbf{U}\,\mathrm{diag}(\mathbf{d})\mathbf{U}^H$.

Proof. The ML cost from Eq. (2.2) is the cost in Eq. (2.7), with $\alpha = n$ and $\mathbf{S} = \mathbf{X}\mathbf{X}^H$. The result then follows immediately from Lemma 2. $\square$

Proposition 3 has been previously shown using duality theory for special cases: $r = p$ and σ^2 known [84]; $r \leq p$ and σ^2 known [2, 15, 53]; $r \leq p$ and $\sigma^2 \geq LB$ [53]. Next, we extend the result to include the persymmetry constraint into the ML estimation.

Proposition 4 (Rank-constrained Persymmetric ML). *Given iid, zero mean, multivariate complex circular Gaussian observations* $\mathbf{X} \in \mathbb{C}^{p \times n}$, $LB \leq \sigma^2 \leq UB$, *and rank* r, $1 \leq r \leq p$. *The maximum likelihood covariance estimate,* $\widehat{\mathbf{R}} = \mathbf{M} + \sigma^2 \mathbf{I}$, *with* $\mathrm{rank}(\mathbf{M}) \leq r$ *and* $\mathbf{M} \succeq 0$ *persymmetric, is* $\widehat{\mathbf{R}} = \mathbf{U}\,\mathrm{diag}(\phi(\mathbf{d}))\mathbf{U}^H$ *with* $\alpha = n$ *and eigendecomposition* $\mathbf{S} = \mathbf{U}\,\mathrm{diag}(\mathbf{d})\mathbf{U}^H$ *for*

$$\mathbf{S} = \tfrac{1}{2}\left\{\mathbf{X}\mathbf{X}^H + \mathbf{J}\left(\mathbf{X}\mathbf{X}^H\right)^T \mathbf{J}\right\}. \tag{2.12}$$

Proof. The proof proceeds in the same manner as the proof of Proposition 2 with $\alpha = n$ and $\mathbf{S}$ as given in Eq. (2.12). $\square$

2.6 Numerical Experiments

For the MAP scenario, the true covariance $\mathbf{R}$ was drawn from f in Section II, which was approximated by drawing iid from $\mathcal{CIW}(\mathbf{R}_0, l = p + 1)$, then projecting each random draw to the set $\mathcal{S} \cap \mathcal{T}$. The scale matrix was $\mathbf{R}_0 = \mathbf{M} + \sigma^2 \mathbf{I}$, with $\mathbf{M}$ constructed from the $r = 5$ largest eigenvectors and corresponding eigenvalues of the exponential tapered model [28], [90] with the i, jth element given as $\rho^{|i-j|}, i, j = 1, \ldots, p$, $\rho = 0.9$, $p = 16$, and $\sigma^2 = 1$ (i.e., LB = UB for known thermal noise [40]). For the ML case, the true covariance was set equal to the scale matrix, $\mathbf{R} = \mathbf{R}_0$.

Eight covariance estimators were computed: the maximum a posteriori (MAP), rank-constrained MAP (RC-MAP, given in Prop. 1), persymmetric MAP (PSM-MAP, given in Eq. (2.11)), rank-constrained persymmetric MAP (RC-PSM-MAP, given in Prop. 2), and their ML counterparts: ML, RC-ML, given in Prop. 3, PSM-ML, given in Eq. (2.12), and RC-PSM-ML, given in Prop. 4, respectively. Of these eight estimators, MAP and ML are unconstrained (i.e., $r = p$, LB = 0, and UB = ∞), and they were computed as $\frac{1}{n+l+p}(n\mathbf{X}\mathbf{X}^H + l\mathbf{R}_0)$ and $\frac{1}{n}\mathbf{X}\mathbf{X}^H$, respectively. For each estimated covariance matrix $\widehat{\mathbf{R}}$, the normalized signal-to-noise ratio (SINR) [47], [53] was computed as $\frac{(\mathbf{s}^H\widehat{\mathbf{R}}^{-1}\mathbf{s})^2}{(\mathbf{s}^H\widehat{\mathbf{R}}^{-1}\mathbf{R}\widehat{\mathbf{R}}^{-1}\mathbf{s})(\mathbf{s}^H\mathbf{R}^{-1}\mathbf{s})}$ with the steering vector $\mathbf{s} = \frac{1}{\sqrt{p}}[1, \cdots, 1]^T$. Results are omitted for the unconstrained ML estimator for $n < p$ because of its singularity. The results were averaged over 10000 Monte-Carlo simulations for each scenario. Normalized SINR loss versus number of snapshots is shown in Fig. 2.1 for the MAP estimator and in Fig. 2.2 for the ML estimator.

As expected, the results show increasing performance with increasing knowledge of the covariance matrix structure. In both the MAP and ML scenarios, the estimators employing persymmetry and/or rank constraints outperform their less constrained

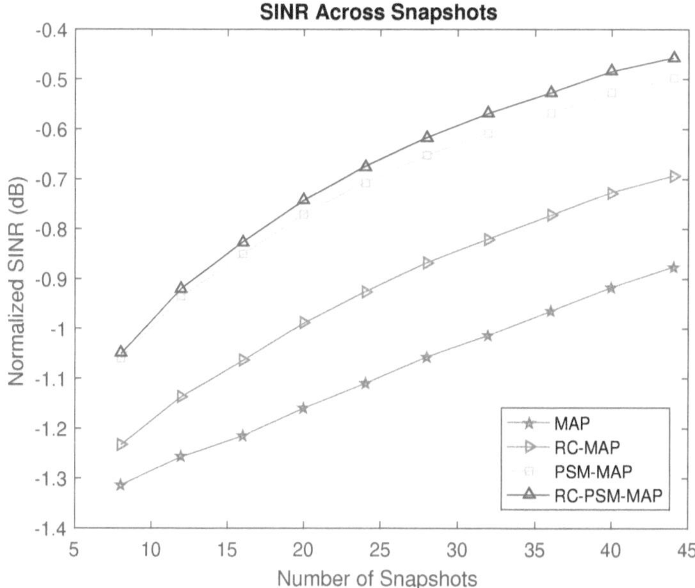

Figure 2.1: Normalized SINR loss versus number of samples, n, for maximum a posteriori covariance matrix estimator.

counterparts, demonstrating the efficacy of the simple, closed-form solutions given. The results further suggest that, given a reasonable prior, the MAP estimate can outperform ML. Modest SINR gains are observed in the ML case without additional computational complexity; significant relative SINR gains are observed in the MAP case. The intuitively expected SINR gain from persymmetry is observed in Fig. 2.1; for example, comparing RC-PSM-MAP to RC-MAP at SINR $= -0.85$ shows a 50% reduction in the required number of snapshots.

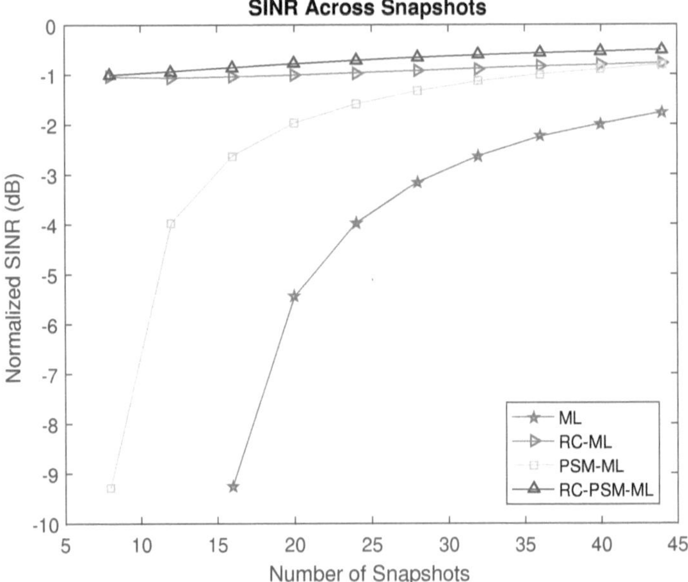

Figure 2.2: Normalized SINR loss versus number of samples, n, for maximum likelihood covariance matrix estimator.

2.7 Conclusion

MAP and ML covariance estimators were extended to low-rank structured covariance matrices. Structure was a low-rank component plus a noise floor, $\mathbf{R} = \mathbf{M} + \sigma^2 \mathbf{I}$; additional structure of persymmetry was also considered. Estimates were obtained as simple, closed-form solutions, despite the non-convexity of the set of low-rank $\mathbf{M}$. Numerical simulations confirmed the intuition that rank and symmetry constraints improve the accuracy of a covariance matrix estimate, and allow for its estimation

even when the number of homogeneous, target-free samples is less than the matrix
dimension.

Chapter 3: Adaptive Detection in the Presence of Partial Homogeneity and Low Condition Number

3.1 Overview

In this chapter, two new detectors are proposed for the partially homogeneous environment by incorporating a fictitious orthogonal component in the null hypothesis and forming a generalized likelihood ratio test for this altered scenario. The motivation for this approach is to improve detection performance in the presence of low training data, akin to what has been achieved by a similar detector for the fully homogeneous scenario. An alternative derivation is also presented for one of the proposed detectors, based on maxi-min SINR linear filtering. Several connections are shown between the proposed detectors and those previously studied in the literature. Moreover, the new detectors are shown to have a Constant False Alarm Rate (CFAR) with respect to both the structure and scaling of the covariance matrix. The presented empirical results suggest that in certain sample-starved scenarios, the proposed detectors achieve higher probabilities of detection than the classical Adaptive Coherence Estimator (ACE), and show a lower degradation in detection performance under the presence of jamming.

3.2 Introduction

The detection of a signal of interest contaminated by Gaussian noise is an essential problem in signal processing, and one that is continuing to generate active research. Of particular interest has been the situation of an adaptive airborne radar system [47], which must detect the presence or absence of a target signal amongst clutter returns. For the point-like target scenario, the signal of interest is commonly modeled as one or more known steering vectors with unknown scaling factors, while the clutter returns and thermal noise (henceforth collectively referred to as noise) are modeled as samples from a zero-mean colored Gaussian distribution with an unknown covariance matrix. A number of secondary training data (a.k.a. snapshots), assumed to be statistically homogeneous and target-free, are required for estimating the unknown covariance matrix and performing the whitened matched filtering operation that is used in the one-step and two-step Generalized Likelihood Ratio (GLR) detectors, known as Kelly's GLR Test (KGLRT) [56] and the Adaptive Matched filter (AMF) [80], respectively.

However, the presence of various inhomogeneities in real-world clutter returns brings forth the dual problems of heterogeneity and scarcity of the secondary data. One approach that offers a degree of robustness against data heterogeneity while maintaining mathematical tractability is the model of a partially homogeneous environment; namely, one in which the covariance matrices of the Cell(s) Under Test (CUT) and the secondary target-free data differ by a positive but unknown scalar constant. With regards to data scarcity, a number of approaches have been proposed in the literature (see, e.g., [37, 84, 29, 28, 90, 5, 53, 6, 79, 83]) wherein various structural constraints and/or statistical priors have been placed on the covariance matrix, whose resulting estimate differs from the Sample Covariance Matrix (SCM), which is the unconstrained Maximum Likelihood (ML) estimate used in GLRT-based detectors. Another technique introduced in [1], called expected likelihood, was used to find an alternative covariance estimate to the ML-based approaches. However, in all of these approaches, the resulting detectors using the alternative covariance estimates can only have a constant (or "approximately constant") false alarm rate (CFAR) when the true covariance shares the same underlying structure as that which is enforced onto the estimate. In contrast to these approaches, a detector has been proposed in [65], called the Adaptive Orthogonal Rejection Detector (AORD), which can be proven to be CFAR regardless of the true covariance matrix structure. Furthermore, in [63], the authors showed that for covariance matrices with relatively low condition numbers (ratio of largest to smallest eigenvalue), the AORD has a considerably higher probability of detection in the low-snapshot regime than the seminal detectors in [56, 80]. This improved performance is achieved by incorporating a component orthogonal to the known steering vector into the hypothesis testing scenario; though originally introduced as a means to combat unknown jamming, this addition allows for a higher probability of detection of the signal of interest even when no jammers are present. As we will show, the GLRT detector for the hypothesis test incorporating such an orthogonal component uses white noise matched filtering. In this paper, we modify the orthogonal modeling approach to apply to the partially homogeneous scenario, and show a similar performance enhancement, compared to the Adaptive Coherence Estimator (ACE) [60], in the low training data scenario. Moreover, we show that the proposed detector, like the ACE, maintains CFAR with respect to both the unknown covariance matrix structure and its scaling.

33

Notation: We use boldface uppercase letters for matrices and boldface lower case letters for vectors; scalars are in non-bold font. We use the superscript $()^H$ to denote conjugate transpose. The notation $\mathcal{CN}(\mathbf{m}, \mathbf{R})$ will denote a complex circular Gaussian distribution with mean vector $\mathbf{m}$ and covariance matrix $\mathbf{R}$, while the notation $\mathcal{CW}(K, \mathbf{R})$ will denote a complex Wishart distribution with K degrees of freedom and scale matrix $\mathbf{R}$; the dimensions for both of these distributions will be clear from the context. We will denote the Frobenius norm of a matrix $\mathbf{M}$ as $\|\mathbf{M}\|_F = \sqrt{\mathrm{tr}\,(\mathbf{M}^H\mathbf{M})}$. For a full-rank matrix $\mathbf{B} \in \mathbb{C}^{N \times M}$, $N > M$, the projection onto the subspace spanned by the columns of $\mathbf{B}$ will be represented as $\mathbf{P_B} = \mathbf{B}(\mathbf{B}^H\mathbf{B})^{-1}\mathbf{B}^H$, while the projection onto the orthogonal complement of the column space will be denoted as $\mathbf{P_B^{\perp}} = \mathbf{I} - \mathbf{P_B}$.

3.3 Proposed Detectors

3.3.1 Problem Setup

The hypothesis test we consider is given by

$$\begin{cases} H_1 : \mathbf{X} = \mathbf{V}\mathbf{\Theta} + \mathbf{N}, & \mathbf{x}_k = \mathbf{n}_k, \ k = 1, 2, \ldots, K_s \\ H_0 : \mathbf{X} = \mathbf{N}, & \mathbf{x}_k = \mathbf{n}_k, \ k = 1, 2, \ldots, K_s, \end{cases} \tag{3.1}$$

where $\mathbf{V} \in \mathbb{C}^{N \times r}$, $1 \leq r < N$, is a known orthonormal matrix spanning the signal subspace, and $\mathbf{\Theta} \in \mathbb{C}^{r \times K_p}$ represents the deterministic, unknown coordinates of the hypothetical signal within this subspace. The matrix $\mathbf{X}$ represents the CUT, while $\mathbf{x_k}$ are the secondary signal-free training data. For the partially homogeneous scenario, the additive noise is modeled such that $\mathbf{n} \sim \mathcal{CN}(\mathbf{0}, \gamma\mathbf{R})$, while $\mathbf{n_k} \sim \mathcal{CN}(\mathbf{0}, \mathbf{R})$, and $\gamma > 0$ is an unknown quantity.

While the signal model in (3.1) is fairly general, it is particularly useful for the case of adaptive radar systems [57], wherein it provides sufficient flexibility to model target detection in various scenarios of interest. For example, $\mathbf{V}$ can be constructed using several steering vectors sampled across the spatial and/or Doppler domains when uncertainties exist about the target signature [11], and can be collapsed down to a single vector $\mathbf{v} \in \mathbb{C}^N$ in the case of $r = 1$ when the target signature is known more precisely. Likewise, the parameter K_p, representing the number of vectors in the CUT which may or may not contain a target signal, can be used to model both point-like targets when $K_p = 1$ and range-spread targets for $K_p > 1$, where the latter may occur when dealing with high-resolution radar systems, or when there may be multiple point-like reflectors in close proximity to one another across range [24]. Finally, the inclusion of the scaling parameter γ between the covariance matrices of the primary and secondary data allows a mathematically tractable way of modeling real-life data heterogeneity which almost invariably occurs in airborne radar systems. While the γ factor allows for robustness to mismatched covariance scaling, it has also been shown that in certain cases, the GLRT detector designed for the partially homogeneous environment also coincides with the GLRT for other forms of mismatch between the primary and secondary data, such as the presence of unknown rank-one interference [10], and the presence of a mismatch wherein the primary and secondary data covariances are only related by an inverse Wishart distribution [14].

In order to derive our detectors, we modify the scenario of (3.1) to be of the form

$$\begin{cases} H_1 : \mathbf{X} = \mathbf{V\Theta} + \mathbf{N}, & \mathbf{x}_k = \mathbf{n}_k, \ k = 1, 2, \dots, K_s \\ H_0 : \mathbf{X} = \mathbf{V}_\perp \mathbf{\Phi} + \mathbf{N}, & \mathbf{x}_k = \mathbf{n}_k, \ k = 1, 2, \dots, K_s, \end{cases} \tag{3.2}$$

where the term $\mathbf{V}_\perp \mathbf{\Phi}$, consisting of the known matrix $\mathbf{V}_\perp \in \mathbb{C}^{N \times N - r}$ with orthonormal vectors such that $\mathbf{V}_\perp^H \mathbf{V} = \mathbf{0}$, and the unknown $\mathbf{\Phi} \in \mathbb{C}^{(N-r) \times K_p}$, represents a *fictitious* signal from a subspace orthogonal to the steering vector. As we will soon explain, our introduction of a fictitious signal into the null hypothesis is similar to what has been done in works like [76] and [7], albeit for different intents and purposes.

A number of remarks are in order with respect to the scenario given in (3.2). Firstly, note that unlike the case for the AORD [65] [63], here the orthogonal component $\mathbf{V}_\perp \mathbf{\Phi}$ has been dropped from the H_1 hypothesis. This stems from the fact that the AORD hypotheses, which are a special case of the hypotheses given in [8, eq. 1], do not admit a GLRT-based detector for the partially homogeneous scenario. Dropping the $\mathbf{V}_\perp \mathbf{\Phi}$ term from H_1 (but keeping it for H_0) allows us to develop a GLRT-based detector which inherits AORD's powerful detection capability in low training data scenarios. The form for H_0 in this setup actually has the greatest influence on the resulting detector, as we will show that even a GLRT developed for the case of $\mathbf{\Theta} = \mathbf{0}$ can give superior performance in certain low training data scenarios. Secondly, the setup in (3.2) differs from those in papers like [76, 48, 51, 17, 64, 62], where the orthogonality in the H_0 hypothesis is either of the form $\mathbf{V}_\perp{}^H \mathbf{S}^{-1} \mathbf{V} = \mathbf{0}$ (orthogonality in the pseudowhitened space, where $\mathbf{S}$ is the SCM), or of the form $\mathbf{V}_\perp{}^H \mathbf{R}^{-1} \mathbf{V} = \mathbf{0}$ (orthogonality in the truly whitened space). While the overarching goal of the detectors like those in [76, 48, 51, 17, 64, 62] is improved rejection (decreased detection probability) for signals with mismatched steering vectors, the main goal of our proposed detector is to provide a higher detection probability for signals with matched steering vectors, particularly in the commonly encountered low training data scenario.

To derive our detectors, we let $\bar{\mathbf{X}}_i, i = 0, 1$, be given as $\bar{\mathbf{X}}_1 = \mathbf{X} - \mathbf{V}\mathbf{\Theta}$ for H_1 and $\bar{\mathbf{X}}_0 = \mathbf{X} - \mathbf{V}_\perp \mathbf{\Phi}$ for H_0. Let $\mathbf{X}_s = [\mathbf{x}_1 \ \ldots \mathbf{x}_{K_s}]$, and let $\mathbf{S} = \mathbf{X}_s \mathbf{X}_s^H$ denote the (unscaled) SCM for the secondary data. Following an approach similar to those shown in [24, 67], we will first derive the one-step GLRT, which treats the covariance matrix $\mathbf{R}$ as unknown, and then the two-step GLRT, which first assumes $\mathbf{R}$ is known, and then replaces it with an ML-optimal estimate found from the secondary data.

3.3.2 One-step GLRTs for $K_p > 1$

The expression to be evaluated for the one-step GLRT is given as

$$\frac{\max_{\boldsymbol{\Theta}} \max_{\gamma} \max_{\mathbf{R}} f_{H_1}(\mathbf{X}, \mathbf{X}_s; \mathbf{R}, \gamma, \boldsymbol{\Theta})}{\max_{\boldsymbol{\Phi}} \max_{\gamma} \max_{\mathbf{R}} f_{H_0}(\mathbf{X}, \mathbf{X}_s; \mathbf{R}, \gamma, \boldsymbol{\Phi})} \underset{H_0}{\overset{H_1}{\gtrless}} \xi. \tag{3.3}$$

The likelihood function for H_1 is given as

$$f_{H_1}(\mathbf{X}, \mathbf{X}_s; \mathbf{R}, \gamma, \boldsymbol{\Theta}) = \frac{1}{\gamma^{NK_p} \det(\mathbf{R})^{K_p+K_s}} \exp(-\operatorname{tr}\{\mathbf{R}^{-1}(\mathbf{S} + \frac{1}{\gamma}\bar{\mathbf{X}}_1\bar{\mathbf{X}}_1^H)\}). \tag{3.4}$$

The ML estimate for $\mathbf{R}$ in this case is well-known to be $\widehat{\mathbf{R}} = \frac{1}{K_p+K_s}(\mathbf{S} + \frac{1}{\gamma}\bar{\mathbf{X}}_1\bar{\mathbf{X}}_1^H)$. Plugging this value in and simplifying, we get

$$f_{H_1}(\mathbf{X}, \mathbf{X}_s; \widehat{\mathbf{R}}, \gamma, \boldsymbol{\Theta}) = \frac{\exp\left(-\frac{N}{K_p+K_s}\right)}{\gamma^{NK_p} \det\left(\mathbf{S} + \frac{1}{\gamma}\bar{\mathbf{X}}_1\bar{\mathbf{X}}_1^H\right)^{K_p+K_s}}$$

$$\propto \frac{1}{\gamma^{NK_p} \det\left(\mathbf{I} + \frac{1}{\gamma}\bar{\mathbf{X}}_1^H\mathbf{S}^{-1}\bar{\mathbf{X}}_1\right)^{K_p+K_s}}, \tag{3.5}$$

where the proportionality follows from the fact that for conformable matrices $\mathbf{A}$, $\mathbf{B}$, and $\mathbf{C}$, $\det(\mathbf{A} + \mathbf{BC}) = \det(\mathbf{A})\det(\mathbf{I} + \mathbf{CA}^{-1}\mathbf{B})$, and that for our purposes $\det(\mathbf{S})$ is simply a constant independent of the CUT. The next step is to maximize the function with respect to $\boldsymbol{\Theta}$ by minimizing the denominator. To proceed, we use the result in Appendix A to write

$$\arg\min_{\boldsymbol{\Theta}} \det\left(\mathbf{I} + \frac{1}{\gamma}\bar{\mathbf{X}}_1^H\mathbf{S}^{-1}\bar{\mathbf{X}}_1\right) = \arg\min_{\boldsymbol{\Theta}} \operatorname{tr}\left(\mathbf{I} + \frac{1}{\gamma}\bar{\mathbf{X}}_1^H\mathbf{S}^{-1}\bar{\mathbf{X}}_1\right) \tag{3.6}$$

$$= \arg\min_{\boldsymbol{\Theta}} \operatorname{tr}\left(\bar{\mathbf{X}}_1^H\mathbf{S}^{-1}\bar{\mathbf{X}}_1\right) = \arg\min_{\boldsymbol{\Theta}} \|\widetilde{\mathbf{X}} - \widetilde{\mathbf{V}}\boldsymbol{\Theta}\|_F^2,$$

where $\widetilde{\mathbf{X}} = \mathbf{S}^{-\frac{1}{2}}\mathbf{X}$, $\widetilde{\mathbf{V}} = \mathbf{S}^{-\frac{1}{2}}\mathbf{V}$, and where we utilized the positive semidefinite nature of the matrix $\bar{\mathbf{X}}_1^H \mathbf{S}^{-1} \bar{\mathbf{X}}_1$. Then the least-squares (and also ML) estimate of $\boldsymbol{\Theta}$ is well-known to be $\widehat{\boldsymbol{\Theta}} = (\mathbf{V}^H \mathbf{S}^{-1} \mathbf{V})^{-1} \mathbf{V}^H \mathbf{S}^{-1} \mathbf{X}$. Putting this estimate back into (3.5) gives

$$f_{H_1}(\mathbf{X}, \mathbf{X}_s; \widehat{\mathbf{R}}, \gamma, \widehat{\boldsymbol{\Theta}}) \propto \frac{1}{\gamma^{NK_p} \det{(\mathbf{I} + \frac{1}{\gamma}\widetilde{\mathbf{X}}^H \widetilde{\mathbf{P}}_{\widetilde{\mathbf{V}}}^{\perp} \widetilde{\mathbf{X}})}^{K_p + K_s}}, \tag{3.7}$$

where $\widetilde{\mathbf{P}}_{\widetilde{\mathbf{V}}}^{\perp} = \mathbf{I} - \mathbf{S}^{-\frac{1}{2}}\mathbf{V}(\mathbf{V}^H \mathbf{S}^{-1} \mathbf{V})^{-1} \mathbf{V}^H \mathbf{S}^{-\frac{1}{2}}$. Noting that

$$\widetilde{\mathbf{P}}_{\widetilde{\mathbf{V}}}^{\perp} = \bar{\mathbf{P}}_{\bar{\mathbf{V}}_{\perp}} = \bar{\mathbf{V}}_{\perp}(\bar{\mathbf{V}}_{\perp}^H \bar{\mathbf{V}}_{\perp})^{-1} \bar{\mathbf{V}}_{\perp}^H, \tag{3.8}$$

since $\widetilde{\mathbf{V}}^H \bar{\mathbf{V}}_{\perp} = \mathbf{0}$ for $\bar{\mathbf{V}}_{\perp} = \mathbf{S}^{\frac{1}{2}}\mathbf{V}_{\perp}$, we can rewrite (3.7) as

$$f_{H_1}(\mathbf{X}, \mathbf{X}_s; \widehat{\mathbf{R}}, \gamma, \widehat{\boldsymbol{\Theta}}) \propto \frac{1}{\gamma^{NK_p} \det{(\mathbf{I} + \frac{1}{\gamma}\mathbf{X}^H \mathbf{V}_{\perp}(\mathbf{V}_{\perp}^H \mathbf{S} \mathbf{V}_{\perp})^{-1} \mathbf{V}_{\perp}^H \mathbf{X})}^{K_p + K_s}}. \tag{3.9}$$

Now all that is left is to maximize the expression in (3.9) with respect to γ. To do this, note that we can write the $(K_p + K_s)$th root of its denominator as

$$\gamma^{\frac{NK_p}{K_p + K_s}} \prod_{i=1}^{n}(1 + \frac{\lambda_i}{\gamma}), \tag{3.10}$$

where $n = \min(K_p, \ N - r)$ is the number of nonzero eigenvalues λ_i of the matrix $\mathbf{X}^H \mathbf{V}_{\perp}(\mathbf{V}_{\perp}^H \mathbf{S} \mathbf{V}_{\perp})^{-1} \mathbf{V}_{\perp}^H \mathbf{X}$. Taking the logarithm of this quantity, and subsequently setting the derivative with respect to γ to zero, yields the expression

$$\frac{NK_p}{K_p + K_s} - \sum_{i=1}^{n} \frac{\lambda_i}{\lambda_i + \gamma} = 0. \tag{3.11}$$

Clearly, when $\frac{NK_p}{K_p + K_s} < n$, there exists a unique, positive value of $\gamma = \widehat{\gamma}_1$ which satisfies (3.11) and hence maximizes the value in (3.9), since the summation term in (3.11) is a monotonic function of $\gamma > 0$ with range $[0, n)$. We now turn to the H_0 case, for which the likelihood function can be written as

$$f_{H_0}(\mathbf{X}, \mathbf{X}_s; \mathbf{R}, \gamma, \boldsymbol{\Phi}) = \frac{1}{\gamma^{NK_p} \det{(\mathbf{R})}^{K_p + K_s}} \exp(-\operatorname{tr}\{\mathbf{R}^{-1}(\mathbf{S} + \frac{1}{\gamma}\bar{\mathbf{X}}_0 \bar{\mathbf{X}}_0^H)\}). \tag{3.12}$$

Following steps analogous to the H_1 case, we can replace $\mathbf{R}$ and $\boldsymbol{\Phi}$ with their ML estimates $\widehat{\mathbf{R}} = \frac{1}{K_p+K_s}(\mathbf{S} + \frac{1}{\gamma}\bar{\mathbf{X}}_0\bar{\mathbf{X}}_0^H)$ and $\widehat{\boldsymbol{\Phi}} = (\mathbf{V}_\perp^H\mathbf{S}^{-1}\mathbf{V}_\perp)^{-1}\mathbf{V}_\perp^H\mathbf{S}^{-1}\mathbf{X}$ and plug them back into (3.12), giving

$$f_{H_0}(\mathbf{X}, \mathbf{X}_s; \widehat{\mathbf{R}}, \gamma, \widehat{\boldsymbol{\Phi}}) \propto \frac{1}{\gamma^{NK_p} \det\left(\mathbf{I} + \frac{1}{\gamma}\widetilde{\mathbf{X}}^H\widetilde{\mathbf{P}}_{\widetilde{\mathbf{V}}_\perp}^\perp \widetilde{\mathbf{X}}\right)^{K_p+K_s}}. \tag{3.13}$$

Noting again that $\widetilde{\mathbf{P}}_{\widetilde{\mathbf{V}}}^\perp = \bar{\mathbf{P}}_{\bar{\mathbf{V}}}$, we can rewrite (3.13) as

$$f_{H_0}(\mathbf{X}, \mathbf{X}_s; \widehat{\mathbf{R}}, \gamma, \widehat{\boldsymbol{\Phi}}) \propto \frac{1}{\gamma^{NK_p} \det\left(\mathbf{I} + \frac{1}{\gamma}\mathbf{X}^H\mathbf{V}(\mathbf{V}^H\mathbf{S}\mathbf{V})^{-1}\mathbf{V}^H\mathbf{X}\right)^{K_p+K_s}}. \tag{3.14}$$

The value $\widehat{\gamma}_0$ which maximizes this expression is the one which satisfies the relationship in (3.11), where now $n = \min(K_p, r)$ and the eigenvalues λ_i are those of the matrix $\mathbf{X}^H\mathbf{V}(\mathbf{V}^H\mathbf{S}\mathbf{V})^{-1}\mathbf{V}^H\mathbf{X}$. Forming the ratio of $f_{H_1}(\mathbf{X}, \mathbf{X}_s; \widehat{\mathbf{R}}, \widehat{\gamma}, \widehat{\boldsymbol{\Theta}})$ and $f_{H_0}(\mathbf{X}, \mathbf{X}_s; \widehat{\mathbf{R}}, \widehat{\gamma}, \widehat{\boldsymbol{\Phi}})$ and taking the $(K_p + K_s)$th root (which doesn't affect the detection performance), the resulting one-step GLRT detector can be written as

$$\frac{\widehat{\gamma}_0^{\frac{NK_p}{K_p+K_s}} \det\left(\mathbf{I} + \frac{1}{\widehat{\gamma}_0}\mathbf{X}^H\mathbf{V}(\mathbf{V}^H\mathbf{S}\mathbf{V})^{-1}\mathbf{V}^H\mathbf{X}\right)}{\widehat{\gamma}_1^{\frac{NK_p}{K_p+K_s}} \det\left(\mathbf{I} + \frac{1}{\widehat{\gamma}_1}\mathbf{X}^H\mathbf{V}_\perp(\mathbf{V}_\perp^H\mathbf{S}\mathbf{V}_\perp)^{-1}\mathbf{V}_\perp^H\mathbf{X}\right)}. \tag{3.15}$$

We will refer to the new detector given by this statistic as the Partially Homogeneous Orthogonal Rejection Detector (PHORD).

Additionally, we will now derive a one-step GLRT detector for the same scenario as in (3.2), but with $\boldsymbol{\Theta} = \mathbf{0}$. The resulting detector will be useful to consider because it outperforms the ACE in certain data-starved scenarios, and also because the two detectors share an interesting geometrical connection. The optimized likelihood function for H_0 remains the same, while for H_1 it becomes

$$f_{H_1}(\mathbf{X}, \mathbf{X}_s; \widehat{\mathbf{R}}, \gamma, \boldsymbol{\Theta} = \mathbf{0}) \propto \frac{1}{\gamma^{NK_p} \det\left(\mathbf{I} + \frac{1}{\gamma}\widetilde{\mathbf{X}}^H\mathbf{S}^{-1}\widetilde{\mathbf{X}}\right)^{K_p+K_s}}. \tag{3.16}$$

The optimal $\widehat{\gamma}_1$ for this case is given by the solution to (3.11) when $n = \min(K_p, N)$ and the λ_i are the eigenvalues of $\widetilde{\mathbf{X}}^H \mathbf{S}^{-1} \widetilde{\mathbf{X}}$. The one-step GLRT statistic then becomes

$$\frac{\widehat{\gamma}_0^{\frac{NK_p}{K_p+K_s}} \det\left(\mathbf{I} + \frac{1}{\widehat{\gamma}_0} \mathbf{X}^H \mathbf{V} (\mathbf{V}^H \mathbf{S} \mathbf{V})^{-1} \mathbf{V}^H \mathbf{X}\right)}{\widehat{\gamma}_1^{\frac{NK_p}{K_p+K_s}} \det\left(\mathbf{I} + \frac{1}{\widehat{\gamma}_1} \mathbf{X}^H \mathbf{S}^{-1} \mathbf{X}\right)}. \tag{3.17}$$

For reasons which we will describe later, we will refer to the detector given by (3.17) as the COlored Signal COsine (COSCO) Detector.

3.3.3 Two-step GLRTs for $K_p > 1$

Now, we will derive the PHORD and COSCO detectors using the two-step GLRT, treating $\mathbf{R}$ as known and replacing it with its ML estimate in the final statistic. The likelihood function for H_1 in this case becomes

$$f_{H_1}(\mathbf{X}, \mathbf{X}_s; \gamma, \boldsymbol{\Theta}) \propto \frac{1}{\gamma^{NK_p}} \exp(-\operatorname{tr}(\frac{1}{\gamma} \bar{\mathbf{X}}_1^H \mathbf{R}^{-1} \bar{\mathbf{X}}_1)), \tag{3.18}$$

where we use the proportionality to ignore known constants which will also appear under H_0 and are unrelated to the CUT. Taking the derivative of (3.18) with respect to γ and setting it equal to zero yields the ML estimate $\widehat{\gamma} = \frac{1}{NK_p} \operatorname{tr}(\bar{\mathbf{X}}_1^H \mathbf{R}^{-1} \bar{\mathbf{X}}_1)$. Putting this back into (3.18) and ignoring constants gives

$$f_{H_1}(\mathbf{X}, \mathbf{X}_s; \widehat{\gamma}, \boldsymbol{\Theta}) \propto \frac{1}{\operatorname{tr}(\bar{\mathbf{X}}_1^H \mathbf{R}^{-1} \bar{\mathbf{X}}_1)^{NK_p}}. \tag{3.19}$$

Noting that $\operatorname{tr}(\bar{\mathbf{X}}_1^H \mathbf{R}^{-1} \bar{\mathbf{X}}_1) = \|\widetilde{\bar{\mathbf{X}}} - \widetilde{\bar{\mathbf{V}}}\boldsymbol{\Theta}\|_F^2$, where $\widetilde{\bar{\mathbf{X}}} = \mathbf{R}^{-\frac{1}{2}}\mathbf{X}$ and $\widetilde{\bar{\mathbf{V}}} = \mathbf{R}^{-\frac{1}{2}}\mathbf{V}$, we can find the least-squares (and ML) estimate of $\boldsymbol{\Theta}$ as $\widehat{\boldsymbol{\Theta}} = (\mathbf{V}^H \mathbf{R}^{-1} \mathbf{V})^{-1} \mathbf{V}^H \mathbf{R}^{-1} \mathbf{X}$. Putting this value into (3.19) and using similar relationships to (3.7)-(3.9), we get

$$f_{H_1}(\mathbf{X}, \mathbf{X}_s; \widehat{\gamma}, \widehat{\boldsymbol{\Theta}}) \propto \frac{1}{\operatorname{tr}\left(\widetilde{\bar{\mathbf{X}}}^H \widetilde{\mathbf{P}}_{\mathbf{V}}^\perp \widetilde{\bar{\mathbf{X}}}\right)^{K_p+K_s}} = \frac{1}{\operatorname{tr}\left(\widetilde{\bar{\mathbf{X}}}^H \bar{\bar{\mathbf{P}}}_{\mathbf{V}_\perp} \widetilde{\bar{\mathbf{X}}}\right)^{K_p+K_s}}$$

$$= \frac{1}{\operatorname{tr}\left(\mathbf{X}^H \mathbf{V}_\perp (\mathbf{V}_\perp^H \mathbf{R} \mathbf{V}_\perp)^{-1} \mathbf{V}_\perp^H \mathbf{X}\right)^{K_p+K_s}}, \tag{3.20}$$

where $\widetilde{\mathbf{P}}_{\mathbf{V}}^{\perp} = \mathbf{I} - \widetilde{\mathbf{V}}(\widetilde{\mathbf{V}}^{H}\widetilde{\mathbf{V}})^{-1}\widetilde{\mathbf{V}}^{H}$ and $\bar{\bar{\mathbf{P}}}_{\mathbf{V}_{\perp}} = \bar{\bar{\mathbf{V}}}_{\perp}(\bar{\bar{\mathbf{V}}}_{\perp}^{H}\bar{\bar{\mathbf{V}}}_{\perp})^{-1}\bar{\bar{\mathbf{V}}}_{\perp}^{H}$, for $\bar{\bar{\mathbf{V}}}_{\perp} = \mathbf{R}^{\frac{1}{2}}\mathbf{V}_{\perp}$.

The likelihood function for H_0 is given as

$$f_{H_0}(\mathbf{X}, \mathbf{X}_s; \gamma, \boldsymbol{\Phi}) \propto \frac{1}{\gamma^{NK_p}} \exp(-\operatorname{tr}(\frac{1}{\gamma}\bar{\mathbf{X}}_0^{H}\mathbf{R}^{-1}\bar{\mathbf{X}}_0)). \tag{3.21}$$

Similarly to the H_1 case, we can find the ML estimates of γ and $\boldsymbol{\Phi}$ as

$$\widehat{\gamma} = \frac{1}{NK_p}\operatorname{tr}(\bar{\mathbf{X}}_0^{H}\mathbf{R}^{-1}\bar{\mathbf{X}}_0) \quad \text{and} \quad \widehat{\boldsymbol{\Phi}} = (\mathbf{V}_{\perp}^{H}\mathbf{R}^{-1}\mathbf{V}_{\perp})^{-1}\mathbf{V}_{\perp}^{H}\mathbf{R}^{-1}\mathbf{X}. \tag{3.22}$$

Putting these values into (3.21) and using the relationships analogous to (3.13)-(3.14), we get

$$f_{H_0}(\mathbf{X}, \mathbf{X}_s; \widehat{\gamma}, \widehat{\boldsymbol{\Phi}}) \propto \frac{1}{\operatorname{tr}(\widetilde{\mathbf{X}}^{H}\widetilde{\mathbf{P}}_{\mathbf{V}_{\perp}}^{\perp}\widetilde{\mathbf{X}})^{K_p+K_s}} = \frac{1}{\operatorname{tr}(\widetilde{\mathbf{X}}^{H}\bar{\bar{\mathbf{P}}}_{\mathbf{V}}\widetilde{\mathbf{X}})^{K_p+K_s}}$$

$$= \frac{1}{\operatorname{tr}(\mathbf{X}^{H}\mathbf{V}(\mathbf{V}^{H}\mathbf{R}\mathbf{V})^{-1}\mathbf{V}^{H}\mathbf{X})^{K_p+K_s}}, \tag{3.23}$$

where $\widetilde{\mathbf{P}}_{\mathbf{V}_{\perp}}^{\perp} = \mathbf{I} - \widetilde{\mathbf{V}}_{\perp}(\widetilde{\mathbf{V}}_{\perp}^{H}\widetilde{\mathbf{V}}_{\perp})^{-1}\widetilde{\mathbf{V}}_{\perp}^{H}$ and $\bar{\bar{\mathbf{P}}}_{\mathbf{V}} = \bar{\bar{\mathbf{V}}}(\bar{\bar{\mathbf{V}}}^{H}\bar{\bar{\mathbf{V}}})^{-1}\bar{\bar{\mathbf{V}}}^{H}$, for $\bar{\bar{\mathbf{V}}} = \mathbf{R}^{\frac{1}{2}}\mathbf{V}$. Forming a ratio of the optimized likelihoods, taking the $(K_p + K_s)$th root, and replacing the unknown $\mathbf{R}$ with $\mathbf{S}$, the resulting two-step PHORD statistic is given as

$$\frac{\operatorname{tr}(\mathbf{X}^{H}\mathbf{V}(\mathbf{V}^{H}\mathbf{S}\mathbf{V})^{-1}\mathbf{V}^{H}\mathbf{X})}{\operatorname{tr}(\mathbf{X}^{H}\mathbf{V}_{\perp}(\mathbf{V}_{\perp}^{H}\mathbf{S}\mathbf{V}_{\perp})^{-1}\mathbf{V}_{\perp}^{H}\mathbf{X})}. \tag{3.24}$$

For the two-step COSCO statistic, optimized likelihood for H_0 remains the same, while for H_1 it becomes

$$f_{H_1}(\mathbf{X}, \mathbf{X}_s; \gamma, \boldsymbol{\Theta} = 0) \propto \frac{1}{\gamma^{NK_p}} \exp(-\operatorname{tr}(\frac{1}{\gamma}\mathbf{X}^{H}\mathbf{R}^{-1}\mathbf{X})). \tag{3.25}$$

Using the ML estimate of γ given by $\widehat{\gamma} = \frac{1}{NK_p}\operatorname{tr}(\mathbf{X}^{H}\mathbf{R}^{-1}\mathbf{X})$, replacing $\mathbf{R}$ with $\mathbf{S}$, and finding the $(K_p + K_s)$th root of the resulting ratio of optimized likelihood functions gives the two-step COSCO statistic as

$$\frac{\operatorname{tr}(\mathbf{X}^{H}\mathbf{V}(\mathbf{V}^{H}\mathbf{S}\mathbf{V})^{-1}\mathbf{V}^{H}\mathbf{X})}{\operatorname{tr}(\mathbf{X}^{H}\mathbf{S}^{-1}\mathbf{X})}. \tag{3.26}$$

3.3.4 One-step and Two-step GLRTs for $K_p = 1$

Let $\bar{\mathbf{x}}_i, i = 0, 1$, be given as $\bar{\mathbf{x}}_1 = \mathbf{x} - \mathbf{V}\boldsymbol{\theta}$ for H_1 and $\bar{\mathbf{x}}_0 = \mathbf{x} - \mathbf{V}_\perp \boldsymbol{\phi}$ for H_0. Then, for the H_1 case when $K_p = 1$, replacing $\mathbf{R}$ and $\boldsymbol{\theta}$ with their ML estimates $\widehat{\mathbf{R}} = \frac{1}{K_s+1}(\mathbf{S} + \frac{1}{\gamma}\bar{\mathbf{x}}_1\bar{\mathbf{x}}_1^H)$ and $\widehat{\boldsymbol{\theta}} = (\mathbf{V}^H\mathbf{S}^{-1}\mathbf{V})^{-1}\mathbf{V}^H\mathbf{S}^{-1}\mathbf{x}$ gives

$$f_{H_1}(\mathbf{x}, \mathbf{X}_s; \widehat{\mathbf{R}}, \gamma, \widehat{\boldsymbol{\theta}}) \propto \frac{1}{\gamma^N(1 + \frac{1}{\gamma}\mathbf{x}^H\mathbf{V}_\perp(\mathbf{V}_\perp^H\mathbf{S}\mathbf{V}_\perp)^{-1}\mathbf{V}_\perp^H\mathbf{x})^{K_s+1}}. \tag{3.27}$$

Taking the derivative of (3.27) with respect to γ, setting to zero, and solving gives the optimal value $\widehat{\gamma} = \frac{K-N+1}{N}\mathbf{x}^H\mathbf{V}_\perp(\mathbf{V}_\perp^H\mathbf{S}\mathbf{V}_\perp)^{-1}\mathbf{V}_\perp^H\mathbf{x}$. Utilizing this value in (3.27) gives

$$f_{H_1}(\mathbf{x}, \mathbf{X}; \widehat{\mathbf{R}}, \widehat{\gamma}, \widehat{\boldsymbol{\theta}}) \propto \frac{1}{(\mathbf{x}^H\mathbf{V}_\perp(\mathbf{V}_\perp^H\mathbf{S}\mathbf{V}_\perp)^{-1}\mathbf{V}_\perp^H\mathbf{x})^N}. \tag{3.28}$$

For H_0, replacing $\mathbf{R}$, $\boldsymbol{\phi}$, and γ in $f_{H_0}(\mathbf{x}, \mathbf{X}; \mathbf{R}, \gamma, \boldsymbol{\phi})$ with their ML estimates $\widehat{\mathbf{R}} = \frac{1}{K_s+1}(\mathbf{S} + \frac{1}{\gamma}\bar{\mathbf{x}}_0\bar{\mathbf{x}}_0^H)$, $\widehat{\boldsymbol{\phi}} = (\mathbf{V}_\perp^H\mathbf{S}^{-1}\mathbf{V}_\perp)^{-1}\mathbf{V}_\perp^H\mathbf{S}^{-1}\mathbf{x}$, and $\widehat{\gamma} = \frac{K-N+1}{N}\mathbf{x}^H\mathbf{V}(\mathbf{V}^H\mathbf{S}\mathbf{V})^{-1}\mathbf{V}^H\mathbf{x}$, yields

$$f_{H_0}(\mathbf{x}, \mathbf{X}; \widehat{\mathbf{R}}, \widehat{\gamma}, \widehat{\boldsymbol{\theta}}) \propto \frac{1}{(\mathbf{x}^H\mathbf{V}(\mathbf{V}^H\mathbf{S}\mathbf{V})^{-1}\mathbf{V}^H\mathbf{x})^N}. \tag{3.29}$$

Forming a ratio of the optimized likelihood functions for H_1 and H_0 and taking its Nth root results in the one-step PHORD statistic for $K_p = 1$, which is given by

$$\frac{\mathbf{x}^H\mathbf{V}(\mathbf{V}^H\mathbf{S}\mathbf{V})^{-1}\mathbf{V}^H\mathbf{x}}{\mathbf{x}^H\mathbf{V}_\perp(\mathbf{V}_\perp^H\mathbf{S}\mathbf{V}_\perp)^{-1}\mathbf{V}_\perp^H\mathbf{x}}. \tag{3.30}$$

To derive the one-step COSCO statistic, an identical process is followed for the H_0 case, while for H_1 the ML estimates of $\mathbf{R}$ and γ become $\widehat{\mathbf{R}} = \frac{1}{K_s+1}(\mathbf{S} + \frac{1}{\gamma}\mathbf{x}\mathbf{x}^H)$ and $\widehat{\gamma} = \frac{K-N+1}{N}\mathbf{x}^H\mathbf{S}^{-1}\mathbf{x}$. Using these values in the likelihood function for H_0, forming a ratio with the optimized likelihood for H_1, and taking the Nth root of this ratio gives the one-step COSCO statistic for $K_p = 1$, which can be written as

$$\frac{\mathbf{x}^H\mathbf{V}(\mathbf{V}^H\mathbf{S}\mathbf{V})^{-1}\mathbf{V}^H\mathbf{x}}{\mathbf{x}^H\mathbf{S}^{-1}\mathbf{x}}. \tag{3.31}$$

Finally, we note that for the two-step derivation of PHORD, which first assumes $\mathbf{R}$ is known, the ML estimates of $\boldsymbol{\theta}$ and γ become $\widehat{\boldsymbol{\theta}} = (\mathbf{V}^H\mathbf{R}^{-1}\mathbf{V})^{-1}\mathbf{V}^H\mathbf{R}^{-1}\mathbf{x}$ and $\widehat{\gamma} = \frac{1}{N}\mathbf{x}^H\mathbf{V}_\perp(\mathbf{V}_\perp^H\mathbf{R}\mathbf{V}_\perp)^{-1}\mathbf{V}_\perp^H\mathbf{x}$ for H_1. On the other hand, for H_0, the relevant estimates of $\boldsymbol{\phi}$ and γ become $\widehat{\boldsymbol{\phi}} = (\mathbf{V}_\perp^H\mathbf{R}^{-1}\mathbf{V}_\perp)^{-1}\mathbf{V}_\perp^H\mathbf{R}^{-1}\mathbf{x}$ and $\widehat{\gamma} = \frac{1}{N}\mathbf{x}^H\mathbf{V}(\mathbf{V}^H\mathbf{R}\mathbf{V})^{-1}\mathbf{V}^H\mathbf{x}$. Forming a ratio of the likelihood functions using these values, taking the Nth root, and replacing $\mathbf{R}$ with $\mathbf{S}$ in the final term gives the same statistic as in (3.30). In a similar manner, for the two-step version of COSCO, which uses $\boldsymbol{\theta} = \mathbf{0}$ and $\widehat{\gamma} = \frac{1}{N}\mathbf{x}^H\mathbf{R}^{-1}\mathbf{x}$ for H_1, performing the same operations on the optimized likelihood functions results in the statistic given in (3.31). Thus, we can see that for $K_p = 1$, the one-step and two-step GLRT detectors coincide for both PHORD and COSCO.

3.3.5 CFAR

We will now show that PHORD and COSCO are CFAR with respect to $\mathbf{R}$ and γ, for the scenario given in (3.1). We will first look at the two-step case for $K_p > 1$, and then extend the results to the one-step case.

Introducing two new matrices $\mathbf{R}_1 = \mathbf{V}^H\mathbf{R}\mathbf{V}$ and $\mathbf{R}_2 = \mathbf{V}_\perp^H\mathbf{R}\mathbf{V}_\perp$, we note that $\frac{1}{\sqrt{\gamma}}\mathbf{V}^H\mathbf{X} \overset{c}{\sim} \mathcal{CN}(\mathbf{0}, \mathbf{R}_1)$ and $\frac{1}{\sqrt{\gamma}}\mathbf{V}_\perp^H\mathbf{X} \overset{c}{\sim} \mathcal{CN}(\mathbf{0}, \mathbf{R}_2)$, where we use the notation "$\overset{c}{\sim}$" to mean "columnwise distributed as." Thus, we can rewrite the two-step PHORD term in (3.24) as

$$
\frac{\operatorname{tr}(\mathbf{X}^H\mathbf{V}(\mathbf{V}^H\mathbf{S}\mathbf{V})^{-1}\mathbf{V}^H\mathbf{X})}{\operatorname{tr}(\mathbf{X}^H\mathbf{V}_\perp(\mathbf{V}_\perp^H\mathbf{S}\mathbf{V}_\perp)^{-1}\mathbf{V}_\perp^H\mathbf{X})} = \frac{\frac{1}{\gamma}\operatorname{tr}(\mathbf{X}^H\mathbf{V}\mathbf{R}_1^{-\frac{1}{2}}(\mathbf{R}_1^{-\frac{1}{2}}\mathbf{V}^H\mathbf{S}\mathbf{V}\mathbf{R}_1^{-\frac{1}{2}})^{-1}\mathbf{R}_1^{-\frac{1}{2}}\mathbf{V}^H\mathbf{X})}{\frac{1}{\gamma}\operatorname{tr}(\mathbf{X}^H\mathbf{V}_\perp\mathbf{R}_2^{-\frac{1}{2}}(\mathbf{R}_2^{-\frac{1}{2}}\mathbf{V}_\perp^H\mathbf{S}\mathbf{V}_\perp\mathbf{R}_2^{-\frac{1}{2}})^{-1}\mathbf{R}_2^{-\frac{1}{2}}\mathbf{V}_\perp^H\mathbf{X})}
$$

$$
= \frac{\operatorname{tr}(\widetilde{\mathbf{X}}_1^H\widetilde{\mathbf{S}}_1^{-1}\widetilde{\mathbf{X}}_1)}{\operatorname{tr}(\widetilde{\mathbf{X}}_2^H\widetilde{\mathbf{S}}_2^{-1}\widetilde{\mathbf{X}}_2)}. \tag{3.32}
$$

From this last term, we can see that $\widetilde{\mathbf{X}}_1 = \frac{1}{\sqrt{\gamma}}\mathbf{R}_1^{-\frac{1}{2}}\mathbf{V}^H\mathbf{X} \overset{c}{\sim} \mathcal{CN}(\mathbf{0}, \mathbf{I})$ and $\widetilde{\mathbf{S}}_1 = \mathbf{R}_1^{-\frac{1}{2}}\mathbf{V}^H\mathbf{S}\mathbf{V}\mathbf{R}_1^{-\frac{1}{2}} \sim \mathcal{CW}(K, \mathbf{I})$. Noting that an analogous argument can be made with respect to $\widetilde{\widetilde{\mathbf{X}}}_2$ and $\widetilde{\widetilde{\mathbf{S}}}_2$, and that none of the distributions of these terms are functions of $\mathbf{R}$ or γ, we can state that the two-step PHORD is CFAR with respect to $\mathbf{R}$ and γ. Using similar steps as in (3.32), we can likewise rewrite the COSCO statistic as

$$\frac{\operatorname{tr}\left(\mathbf{X}^H \mathbf{V}(\mathbf{V}^H \mathbf{S}\mathbf{V})^{-1}\mathbf{V}^H \mathbf{X}\right)}{\operatorname{tr}\left(\mathbf{X}^H \mathbf{S}^{-1}\mathbf{X}\right)} = \frac{\operatorname{tr}\left(\widetilde{\mathbf{X}}_1^{\,H}\widetilde{\mathbf{S}}_1^{\,-1}\widetilde{\mathbf{X}}_1\right)}{\operatorname{tr}\left(\widetilde{\mathbf{X}}^{\,H}\widetilde{\mathbf{S}}^{\,-1}\widetilde{\mathbf{X}}\right)}, \tag{3.33}$$

where $\widetilde{\widetilde{\mathbf{X}}} = \frac{1}{\sqrt{\gamma}}\mathbf{R}^{-\frac{1}{2}}\mathbf{X} \stackrel{c}{\sim} \mathcal{CN}(\mathbf{0},\mathbf{I})$ and $\widetilde{\widetilde{\mathbf{S}}} = \mathbf{R}^{-\frac{1}{2}}\mathbf{S}\mathbf{R}^{-\frac{1}{2}} \sim \mathcal{CW}(K,\mathbf{I})$. Thus, we can say that the two-step COSCO detector also shares the CFAR property of the two-step PHORD. Extending this approach to the one-step version of PHORD given in (3.15), we can write

$$\frac{\widehat{\gamma}_0^{\frac{NK_p}{K_p+K_s}}\det\left(\mathbf{I}+\frac{1}{\widehat{\gamma}_0}\mathbf{X}^H\mathbf{V}(\mathbf{V}^H\mathbf{S}\mathbf{V})^{-1}\mathbf{V}^H\mathbf{X}\right)}{\widehat{\gamma}_1^{\frac{NK_p}{K_p+K_s}}\det\left(\mathbf{I}+\frac{1}{\widehat{\gamma}_1}\mathbf{X}^H\mathbf{V}_{\perp}(\mathbf{V}_{\perp}^H\mathbf{S}\mathbf{V}_{\perp})^{-1}\mathbf{V}_{\perp}^H\mathbf{X}\right)} = \frac{\widetilde{\gamma}_0^{\frac{NK_p}{K_p+K_s}}\det\left(\mathbf{I}+\frac{1}{\widetilde{\gamma}_0}\widetilde{\mathbf{X}}_1^{\,H}\widetilde{\mathbf{S}}_1^{\,-1}\widetilde{\mathbf{X}}_1\right)}{\widetilde{\gamma}_1^{\frac{NK_p}{K_p+K_s}}\det\left(\mathbf{I}+\frac{1}{\widetilde{\gamma}_1}\widetilde{\mathbf{X}}_2^{\,H}\widetilde{\mathbf{S}}_2^{\,-1}\widetilde{\mathbf{X}}_2\right)}, \tag{3.34}$$

where we now have $\widetilde{\gamma}_0 = \frac{\widehat{\gamma}_0}{\gamma}$, and similarly with $\widetilde{\gamma}_1$. The former is now chosen to satisfy the equation

$$\frac{NK_p}{K_p+K_s} - \sum_{i=1}^{n}\frac{\widetilde{\lambda}_i}{\widetilde{\lambda}_i+\widetilde{\gamma}} = 0, \tag{3.35}$$

where $n = \min(K_p, r)$, $\widetilde{\lambda}_i = \frac{\lambda_i}{\gamma}$ are the eigenvalues of the white Wishart matrix $\widetilde{\widetilde{\mathbf{S}}}_1$, and λ_i are the eigenvalues of $\mathbf{X}^H\mathbf{V}(\mathbf{V}^H\mathbf{S}\mathbf{V})^{-1}\mathbf{V}^H\mathbf{X} \sim \mathcal{CW}(K,\gamma\mathbf{I})$. The value $\widetilde{\gamma}_1$ is also chosen to satisfy (3.35), this time with $n = \min(K_p, N-r)$ and $\widetilde{\lambda}_i$ being the eigenvalues of $\widetilde{\widetilde{\mathbf{S}}}_2$. Similarly to the two-step case, we can say that the one-step PHORD is CFAR with respect to $\mathbf{R}$ and γ since its detection statistic in (3.34) is simply a function of the eigenvalues of white Wishart matrices, whose distributions are independent of these two quantities. Finally, we note that the technique in (3.34) can be analogously extended to the one-step COSCO case by writing

$$\frac{\widehat{\gamma}_0^{\frac{NK_p}{K_p+K_s}} \det\left(\mathbf{I} + \frac{1}{\widehat{\gamma}_0}\mathbf{X}^H\mathbf{V}(\mathbf{V}^H\mathbf{S}\mathbf{V})^{-1}\mathbf{V}^H\mathbf{X}\right)}{\widehat{\gamma}_1^{\frac{NK_p}{K_p+K_s}} \det\left(\mathbf{I} + \frac{1}{\widehat{\gamma}_1}\mathbf{X}^H\mathbf{S}^{-1}\mathbf{X}\right)} = \frac{\widetilde{\gamma}_0^{\frac{NK_p}{K_p+K_s}} \det\left(\mathbf{I} + \frac{1}{\widetilde{\gamma}_0}\widetilde{\mathbf{X}}_1^H\widetilde{\mathbf{S}}_1^{-1}\widetilde{\mathbf{X}}_1\right)}{\widetilde{\gamma}_1^{\frac{NK_p}{K_p+K_s}} \det\left(\mathbf{I} + \frac{1}{\widetilde{\gamma}_1}\widetilde{\mathbf{X}}^H\widetilde{\mathbf{S}}^{-1}\widetilde{\mathbf{X}}\right)}.$$

Similar arguments as in the PHORD case can be made to show the CFARness of the one-step COSCO detector. These arguments can also be straightforwardly extended to the $K_p = 1$ case, wherein the one-step and two-step PHORD and COSCO detectors coincide, as shown earlier.

3.4 Some Comments on the Proposed and Related Detectors

3.4.1 Connections to Cell-Averaging CFAR Detectors

Several remarks are in order regarding the PHORD statistic in (3.30) and the COSCO statistic in (3.31); the discussion in this section will be focused on the case when $r = 1$ and $K_p = 1$, meaning that the target signal to be detected is a single known steering vector in one CUT with an unknown amplitude. In this case, the numerator of PHORD in (3.30) becomes exactly the AORD statistic given in [65, 63], which can be written as

$$\frac{|\mathbf{v}^H\mathbf{x}|^2}{\mathbf{v}^H\mathbf{S}\mathbf{v}}. \tag{3.36}$$

From this new form, we can see that AORD effectively performs white noise matched filtering on the CUT, and only uses the SCM as a scaling term in the denominator to ensure CFAR. Here, a connection can be made to the AMF [80], whose detection statistic can be written as [2]

$$\frac{|\mathbf{v}^H\mathbf{S}^{-1}\mathbf{x}|^2}{\mathbf{v}^H\mathbf{S}^{-1}\mathbf{v}} \underset{H_0}{\overset{H_1}{\gtrless}} \xi_{AMF} \iff |\mathbf{v}^H\mathbf{S}^{-1}\mathbf{x}|^2 \underset{H_0}{\overset{H_1}{\gtrless}} \xi_{AMF}\mathbf{v}^H\mathbf{S}^{-1}\mathbf{v}. \tag{3.37}$$

---

[2]The original AMF statistic was given using the scaled SCM $\frac{1}{K}\mathbf{S}$; here, we ignore this scaling for simplicity and assume that the threshold ξ_{AMF} has absorbed this scaling difference.

As shown in [80], the right-most value in (3.37) can be expanded as

$$\mathbf{v}^H\mathbf{S}^{-1}\mathbf{v} = \mathbf{v}^H\mathbf{S}^{-1}\mathbf{S}\mathbf{S}^{-1}\mathbf{v} = \sum_{k=1}^{K}\mathbf{v}^H\mathbf{S}^{-1}\mathbf{x}_k\mathbf{x}_k^H\mathbf{S}^{-1}\mathbf{v} = \sum_{k=1}^{K}|\widetilde{\mathbf{v}}^H\widetilde{\mathbf{x}}_k|^2, \tag{3.38}$$

where $\widetilde{\mathbf{v}} = \mathbf{S}^{-\frac{1}{2}}\mathbf{v}$ and $\widetilde{\mathbf{x}}_k = \mathbf{S}^{-\frac{1}{2}}\mathbf{x}_k$. Similarly, the denominator of AORD can be expanded as

$$\mathbf{v}^H\mathbf{S}\mathbf{v} = \sum_{k=1}^{K}\mathbf{v}^H\mathbf{x}_k\mathbf{x}_k^H\mathbf{v} = \sum_{k=1}^{K}|\mathbf{v}^H\mathbf{x}_k|^2. \tag{3.39}$$

The likelihood ratio tests using the two statistics can now be written in a similar form, with the one for AMF given as

$$|\widetilde{\mathbf{v}}^H\widetilde{\mathbf{x}}|^2 \underset{H_0}{\overset{H_1}{\gtrless}} \xi_{AMF} \sum_{k=1}^{K}|\widetilde{\mathbf{v}}^H\widetilde{\mathbf{x}}_k|^2, \tag{3.40}$$

and for the AORD as

$$|\mathbf{v}^H\mathbf{x}|^2 \underset{H_0}{\overset{H_1}{\gtrless}} \xi_{AORD} \sum_{k=1}^{K}|\mathbf{v}^H\mathbf{x}_k|^2. \tag{3.41}$$

From (3.40) and (3.41), we can interpret both the AMF and the AORD as cell-averaging CFAR (CACFAR) detectors, where the AMF operates in the pseudo-whitened domain, and the AORD in the original colored one. The relevance of this important observation to our case is that the proposed PHORD and COSCO are scaled versions of the AORD, while the classical ACE is a scaled version of the AMF, and these detectors can be written as

$$t_{PHORD} = \frac{|\mathbf{v}^H\mathbf{x}|^2}{(\mathbf{v}^H\mathbf{S}\mathbf{v})(\mathbf{x}^H\mathbf{V}_\perp(\mathbf{V}_\perp^H\mathbf{S}\mathbf{V}_\perp)^{-1}\mathbf{V}_\perp^H\mathbf{x})}, \tag{3.42}$$

$$t_{COSCO} = \frac{|\mathbf{v}^H\mathbf{x}|^2}{(\mathbf{v}^H\mathbf{S}\mathbf{v})(\mathbf{x}^H\mathbf{S}^{-1}\mathbf{x})}, \tag{3.43}$$

and

$$t_{ACE} = \frac{|\mathbf{v}^H\mathbf{S}^{-1}\mathbf{x}|^2}{(\mathbf{v}^H\mathbf{S}^{-1}\mathbf{v})(\mathbf{x}^H\mathbf{S}^{-1}\mathbf{x})}. \tag{3.44}$$

The use of white noise matched filtering brings an interesting trade-off into play. On the one hand, it is well-known that colored noise matched filtering is SINR-optimal, which is significant because the probabilities of detection of most detectors monotonically increase with post-filtered SINR. When the true covariance matrix $\mathbf{R}$ of the interference and noise is known, the SINR improvement of the colored noise matched filter over the white noise one can be described by the Kantorovich inequality [1, 71], given as

$$(\mathbf{v}^H \mathbf{R} \mathbf{v})(\mathbf{v}^H \mathbf{R}^{-1} \mathbf{v}) \leq \frac{(\lambda_1 + \lambda_N)^2}{4\lambda_1 \lambda_N}, \tag{3.45}$$

where λ_1 and λ_N are the largest and smallest of the N eigenvalues of $\mathbf{R}$. The value of the right hand side of (3.45) increases proportionally with the condition number $\frac{\lambda_1}{\lambda_N}$. On the other hand, when the true $\mathbf{R}$ is unknown, and is instead replaced with the SCM estimate $\mathbf{S}$, the degree of improvement in SINR becomes more complicated. This is due to the fact that using $\mathbf{S}$ in the colored noise matched filtering operation gives both an improvement due to increased noise suppression, but also a drawback due to the fact that the estimate $\mathbf{S}$ itself can be quite poor in low training data scenarios. In numerical simulations for the fully homogeneous case not shown here, we have observed that for interference and noise covariances with large condition numbers (e.g., > 10000), the AMF achieves a higher probability of detection than the AORD even in low training data scenarios. This is due to the fact that with sufficiently high condition numbers, the SINR increase from colored noise matched filtering exceeds that of white noise matched filtering even when using a poor covariance estimate $\mathbf{S}$. However, when the condition number of the interference and noise covariance is fairly low (e.g., on the order of several hundreds), the AORD has a significantly higher probability of detection than the AMF for low training data scenarios. This was indeed the case as shown in [63], where the authors modeled the interference and noise covariance matrix as being exponentially tapered with a one-lag correlation, i.e. of the form $[\mathbf{R}]_{i,j} = \rho^{|i-j|}$, $0 < \rho < 1$, which is a modeling choice that has been very widely used by other authors as well in the radar detection literature (see, e.g., [23, 8, 7, 12, 13, 46, 90, 39, 21], and many others). Heuristically, it can be argued that in this scenario the improved performance of AORD comes from not having to utilize the poor covariance estimate $\mathbf{S}$ in the pseudowhitening operation, since the results in [63] suggest that with a low condition number the suboptimal white noise matched filter can give a higher output SINR than the colored noise matched filter, which is SINR-optimal asymptotically but not finitely (in terms of the training data used to estimate $\mathbf{S}$). As we will further observe in a later section, a similar phenomenon occurs between PHORD/COSCO and ACE, since the former and the latter are equivalent to the AORD and the AMF with additional scalings to provide CFAR in the partially homogeneous environment.

Now, in light of this discussion, we would also like to note that the white noise matched filter employed by the AORD, PHORD, and COSCO can still be SINR-optimal for a special case of colored noise wherein the covariance is the sum of a rank-one target steering vector component and an identity matrix, as we will see in the next subsection.

3.4.2 Derivation as a Maxi-min-Optimal Linear Filter

The optimality of the matched filtering operation $|\mathbf{v}^H \mathbf{x}|^2$ in the presence of white Gaussian noise is well-known. In this section, we will show that this operation is also optimal in the maxi-min sense for colored Gaussian noise; in other words, it maximizes the worst-case SINR. Incorporating appropriate scalings to guarantee CFAR will lead us to an alternative derivation of the AORD and PHORD given in (3.36) and (3.42), respectively.

Let us consider the scenario in (3.1) with $\gamma = 1$, $r = 1$, $K_p = 1$, and $\mathbf{R} = \mathbf{C} + \sigma^2 \mathbf{I}$. If we also let $\theta = 1$, without loss of generality, we have the optimization of the SINR $\frac{\mathrm{E}[|\mathbf{w}^H\mathbf{v}|^2]}{\mathrm{E}[|\mathbf{w}^H\mathbf{n}|^2]} = \frac{|\mathbf{w}^H\mathbf{v}|^2}{\mathbf{w}^H\mathbf{R}\mathbf{w}}$ given as

$$\arg\max_{\mathbf{w}} \min_{\mathbf{C}} \quad \frac{|\mathbf{w}^H\mathbf{v}|^2}{\mathbf{w}^H\mathbf{C}\mathbf{w} + \sigma^2\|\mathbf{w}\|_2^2}$$
$$\text{s. t.} \quad \mathbf{C} \succeq 0$$
$$\mathrm{tr}\,(\mathbf{C}) = 1$$
$$\|\mathbf{w}\|_2 = 1, \tag{3.46}$$

where we assume $\|\mathbf{v}\|_2 = 1$. Using Lemma 1 in [83], which is an extension of [72, 88], we note that

$$\mathbf{w}^H\mathbf{C}\mathbf{w} = \mathrm{tr}\,\left(\mathbf{C}\mathbf{w}\mathbf{w}^H\right) = \mathrm{tr}\,(\mathbf{C}\mathbf{W}) \leq \boldsymbol{\lambda}_{\mathbf{C}}^T \boldsymbol{\lambda}_{\mathbf{W}}, \tag{3.47}$$

where $\boldsymbol{\lambda}_{\mathbf{C}}$ and $\boldsymbol{\lambda}_{\mathbf{W}}$ contain the eigenvalues, in descending order, of the matrices $\mathbf{C}$ and $\mathbf{W} = \mathbf{w}\mathbf{w}^H$, respectively, and where the upper bound is met if and only if $\mathbf{C}$ and $\mathbf{W}$ share the same eigenvectors. Clearly, the choice of $\mathbf{C}$ which meets this upper bound (and hence minimizes the SINR, for the first stage of the optimization in (3.46)) is given by $\mathbf{C}_{\mathrm{opt}} = \mathbf{w}\mathbf{w}^H$. Putting this value back in and simplifying, we are left with

$$\arg\max_{\mathbf{w}} \quad \frac{|\mathbf{w}^H\mathbf{v}|^2}{\|\mathbf{w}\|_2^2(\|\mathbf{w}\|_2^2 + \sigma^2)}$$
$$\text{s. t.} \quad \|\mathbf{w}\|_2 = 1. \tag{3.48}$$

The denominator of (3.48) is a constant, and the numerator is maximized when $\mathbf{w} = \mathbf{v}$ by the Cauchy-Schwarz inequality. Thus, we see that the white noise matched filtering operation $|\mathbf{v}^H\mathbf{x}|^2$ is also optimal in colored noise for the worst-SINR-case scenario when $\mathbf{R} = \mathbf{v}\mathbf{v}^H + \sigma^2\mathbf{I}$, meaning when the colored portion of the noise covariance (corresponding to $\mathbf{C}$) lies directly along the desired steering vector $\mathbf{v}$.

In practice, when using the filter $\mathbf{w} = \mathbf{v}$ for the case $H_0 : \mathbf{x} = \mathbf{n}$, $\mathbf{n} \sim \mathcal{CN}(\mathbf{0}, \mathbf{R})$ for arbitrary $\mathbf{R}$, the filter output is equivalent in distribution to $|(\sqrt{\mathbf{v}^H\mathbf{R}\mathbf{v}})\mathbf{v}^H\breve{\mathbf{x}}|^2$, where $\breve{\mathbf{x}} \sim \mathcal{CN}(\mathbf{0}, \mathbf{I})$. To remove the dependency on $\mathbf{R}$, we can utilize the scaled filter vector $\mathbf{w}_s = \frac{1}{\sqrt{\mathbf{v}^H\mathbf{S}\mathbf{v}}}\mathbf{v}$, where $\sqrt{\mathbf{v}^H\mathbf{S}\mathbf{v}}$ is the unbiased (up to a constant) sample estimate of $\sqrt{\mathbf{v}^H\mathbf{R}\mathbf{v}}$. The output of this filter, $|\mathbf{w}_s^H\mathbf{x}|^2$, is exactly the AORD statistic of (3.36) which is known to be CFAR [63].

To maintain CFAR for the partially homogeneous scenario in (3.1), we must incorporate an additional scaling on $\mathbf{w}$ to account for γ. Here, the only data sample we can use to estimate γ is the CUT itself, $\mathbf{x}$, whose mean is nonzero under H_1. On the other hand, multiplying $\mathbf{x}$ by $\mathbf{V}_\perp^H$ gives a value equivalent in distribution to $\sqrt{\gamma}\mathbf{R}_2^{\frac{1}{2}}\breve{\mathbf{x}}$, where $\mathbf{R}_2 = \mathbf{V}_\perp^H\mathbf{R}\mathbf{V}_\perp$ and $\breve{\mathbf{x}}$ is now zero-mean white Gaussian with length $N-1$. To remove the dependency on $\mathbf{R}$, we can further multiply $\mathbf{V}_\perp^H\mathbf{x}$ by $\mathbf{S}_2^{-\frac{1}{2}}$, where $\mathbf{S}_2 = \mathbf{V}_\perp^H\mathbf{S}\mathbf{V}_\perp$ is the unbiased (up to a constant) sample estimate of $\mathbf{R}_2$. The output $|\mathbf{w}_s^H\mathbf{x}|^2$ using the new scaled filter $\mathbf{w}_s = \frac{1}{\sqrt{(\mathbf{v}^H\mathbf{S}\mathbf{v})(\mathbf{x}^H\mathbf{V}_\perp\mathbf{S}_2^{-1}\mathbf{V}_\perp^H\mathbf{x})}}\mathbf{v}$ is equivalent to the PHORD statistic in (3.42).

3.4.3 Geometrical Interpretations

Now, for the case of $r = 1$ and $K_p = 1$, several connections can be made between COSCO, PHORD, and ACE, which is the GLRT detector [59] for the scenario in (3.1), and whose detection statistic is given in (3.44). Comparing this with the COSCO statistic in (3.43), we can write both as

$$t_{COSCO} = \cos^2(\widetilde{\mathbf{x}}, \bar{\mathbf{v}}) = \frac{|\widetilde{\mathbf{x}}^H\bar{\mathbf{v}}|^2}{\|\widetilde{\mathbf{x}}\|^2\|\bar{\mathbf{v}}\|^2} \tag{3.49}$$

and

$$t_{ACE} = \cos^2(\widetilde{\mathbf{x}}, \widetilde{\mathbf{v}}) = \frac{|\widetilde{\mathbf{x}}^H \widetilde{\mathbf{v}}|^2}{\|\widetilde{\mathbf{x}}\|^2 \|\widetilde{\mathbf{v}}\|^2}. \tag{3.50}$$

Thus, we see that the COSCO and the ACE are the squared cosines of the angle between the pseudowhitened CUT $\widetilde{\mathbf{x}} = \mathbf{S}^{-1}\mathbf{x}$ and either the pseudocolored steering vector $\bar{\mathbf{v}} = \mathbf{S}^{\frac{1}{2}}\mathbf{v}$ (for COSCO) or the pseudowhitened one $\widetilde{\mathbf{v}} = \mathbf{S}^{-\frac{1}{2}}\mathbf{v}$ (for ACE). The implication of this for the detection problem is that while both the COSCO and the ACE form cones of invariances along which the detection performance remains unperturbed, as depicted in Fig. 3.1, COSCO is invariant to rotations around the subspace $\langle\bar{\mathbf{v}}\rangle$, while the ACE is instead invariant to rotations around $\langle\widetilde{\mathbf{v}}\rangle$.

Some comparisons can also be made between the ACE and the PHORD. First, we can write the ACE in an alternative form [59] as

$$\frac{\mathbf{x}^H \mathbf{S}^{-1} \mathbf{x}}{\mathbf{x}^H \mathbf{S}^{-1} \mathbf{x} - \frac{|\mathbf{v}^H \mathbf{S}^{-1} \mathbf{x}|^2}{\mathbf{v}^H \mathbf{S}^{-1} \mathbf{v}}} = \frac{\mathbf{x}^H \mathbf{S}^{-1} \mathbf{x}}{\mathbf{x}^H \mathbf{V}_\perp (\mathbf{V}_\perp^H \mathbf{S} \mathbf{V}_\perp)^{-1} \mathbf{V}_\perp^H \mathbf{x}}, \tag{3.51}$$

where the equivalence between the denominator terms is of the same kind as in (3.7)-(3.9) when $r = 1$. In contrast with the PHORD statistic in (3.42), it can be seen from (3.51) that ACE compares the value $\mathbf{x}^H \mathbf{V}_\perp (\mathbf{V}_\perp^H \mathbf{S} \mathbf{V}_\perp)^{-1} \mathbf{V}_\perp^H \mathbf{x}$ with all the energy present in the pseudowhitened CUT, i.e. $\mathbf{x}^H \mathbf{S}^{-1} \mathbf{x} = \|\widetilde{\mathbf{x}}\|_2^2$, while PHORD compares this value only with that portion of the (scaled) CUT energy that lies along $\mathbf{v}$. A second interpretation can be obtained by expressing the ACE statistic as

$$\frac{\mathbf{x}^H \mathbf{S}^{-1} \mathbf{x}}{\mathbf{x}^H \mathbf{V}_\perp (\mathbf{V}_\perp^H \mathbf{S} \mathbf{V}_\perp)^{-1} \mathbf{V}_\perp^H \mathbf{x}} = \frac{\mathbf{x}^H \mathbf{V}_\perp (\mathbf{V}_\perp^H \mathbf{S} \mathbf{V}_\perp)^{-1} \mathbf{V}_\perp^H \mathbf{x} + \frac{|\mathbf{v}^H \mathbf{S}^{-1} \mathbf{x}|^2}{\mathbf{v}^H \mathbf{S}^{-1} \mathbf{v}}}{\mathbf{x}^H \mathbf{V}_\perp (\mathbf{V}_\perp^H \mathbf{S} \mathbf{V}_\perp)^{-1} \mathbf{V}_\perp^H \mathbf{x}}$$

$$= 1 + \frac{\frac{|\widetilde{\mathbf{v}}^H \widetilde{\mathbf{x}}|^2}{\widetilde{\mathbf{v}}^H \widetilde{\mathbf{v}}}}{\widetilde{\mathbf{x}}^H \widetilde{\mathbf{V}}_\perp (\widetilde{\mathbf{V}}_\perp^H \widetilde{\mathbf{V}}_\perp)^{-1} \widetilde{\mathbf{V}}_\perp^H \widetilde{\mathbf{x}}} \stackrel{m}{\propto} \frac{\|\widetilde{\mathbf{P}}_{\mathbf{v}} \widetilde{\mathbf{x}}\|_2^2}{\|\bar{\mathbf{P}}_{\mathbf{V}_\perp} \widetilde{\mathbf{x}}\|_2^2}. \tag{3.52}$$

For PHORD, the corresponding expression is given in the form

$$\frac{\frac{|\mathbf{v}^H \mathbf{x}|^2}{\mathbf{v}^H \mathbf{S} \mathbf{v}}}{\mathbf{x}^H \mathbf{V}_\perp (\mathbf{V}_\perp^H \mathbf{S} \mathbf{V}_\perp)^{-1} \mathbf{V}_\perp^H \mathbf{x}} = \frac{\frac{|\bar{\mathbf{v}}^H \widetilde{\mathbf{x}}|^2}{\bar{\mathbf{v}}^H \bar{\mathbf{v}}}}{\widetilde{\mathbf{x}}^H \bar{\mathbf{V}}_\perp (\bar{\mathbf{V}}_\perp^H \bar{\mathbf{V}}_\perp)^{-1} \bar{\mathbf{V}}_\perp^H \widetilde{\mathbf{x}}} = \frac{\|\bar{\mathbf{P}}_{\mathbf{v}} \widetilde{\mathbf{x}}\|_2^2}{\|\bar{\mathbf{P}}_{\mathbf{V}_\perp} \widetilde{\mathbf{x}}\|_2^2}. \tag{3.53}$$

Since $\bar{\mathbf{V}}_{\perp}^{H}\widetilde{\mathbf{v}} = \mathbf{0}$, we can see that ACE measures the energy from the pseudowhitened CUT $\widetilde{\mathbf{x}}$ projected onto two orthogonal subspaces and forms a ratio between these projected values. However, since $\bar{\mathbf{V}}_{\perp}^{H}\bar{\mathbf{v}} \neq \mathbf{0}$, we observe that PHORD instead forms a ratio of the energies of $\widetilde{\mathbf{x}}$ projected onto two nonorthogonal subspaces. The implication of this is that for ACE, the test vector $\widetilde{\mathbf{x}}$ can be rotated around the subspace $\langle\widetilde{\mathbf{v}}\rangle$ without changing the value of its projection along $\bar{\mathbf{V}}_{\perp}$. On the other hand, no such rotation can be done for the test vector in PHORD, because it would clearly change the value of its projection along $\bar{\mathbf{V}}_{\perp}$. Thus, while both detectors are invariant with respect to an arbitrary scaling of $\widetilde{\mathbf{x}}$, a necessary condition to maintain CFAR in a partially homogeneous environment, ACE possesses an extra degree of rotational invariance that is absent from PHORD.

3.5 Numerical Experiments

The performance of PHORD and COSCO was compared with that of the ACE [60], the classical detector for the partially homogeneous environment, which is the GLRT for the scenario in (3.1). The experimental signal model also followed (3.1), with γ chosen in each Monte-Carlo (MC) simulation as a uniformly distributed random variable between one and three. For the range-distributed versions, we used the one-step and two-step formulations of ACE as described in [24, 67]. Following similar modeling choices in [65, 63, 7], and [67], the covariance matrix $\mathbf{R}$ was set so that its i, j th element was given as $\rho^{|i-j|}$, with $\rho = 0.9$, each steering vector $\mathbf{v}$ of length $N = 40$ was set as $\mathbf{v} = [1 \ \exp\{-2\pi i\omega\} \ \ldots \ \exp\{-2\pi i(N-1)\omega\}]^{H}$, where i is the imaginary unit and $\omega = 0.1$, and was scaled to have unit norm. For the subspace signal case with $r = 9$, we formed $\mathbf{V}$ from orthonormalized steering vectors at angles

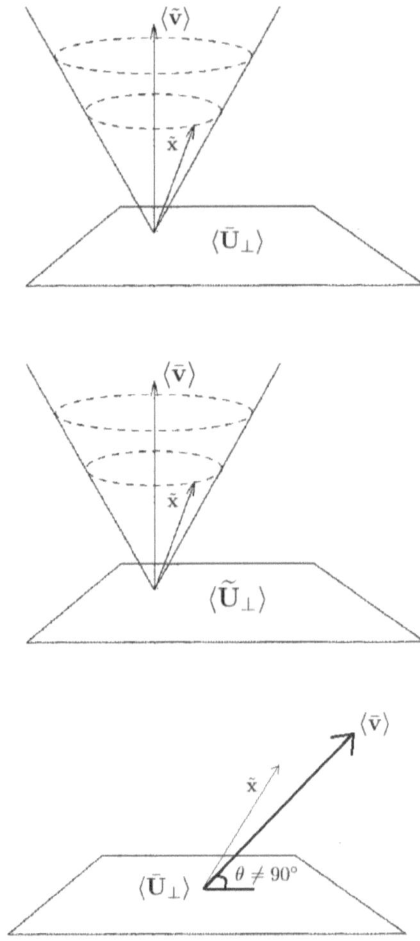

Figure 3.1: Geometrical interpretations, from top to bottom, of the ACE, COSCO, and PHORD (adapted from [60]).

equally spaced from $\omega = 0.1$ to $\omega = 0.3$, while for $r = 1$, the sole steering vector was for $\omega = 0.1$. Each element of the variable $\Theta \in \mathbb{C}^{r \times K_p}$ was chosen as $\mathcal{CN}(0,1)$, and Θ was multiplied by a constant scaling set to achieve a desired SINR, defined as $\frac{1}{K_p} \text{tr} \left(\Theta^H \mathbf{V}^H \mathbf{R}^{-1} \mathbf{V} \Theta \right)$. The false alarm rate, Pfa, was set as 10^{-2}, and the thresholds to be used were set using 5×10^4 MC simulations, while 5×10^3 MC simulations were used for calculating the probability of detection across SINR.

Figures 3.2 and 3.3 show the case when the target signal is confined to one range-cell so that $K_p = 1$, i.e. the point-like case. In the subspace signal case of Fig. 3.2, PHORD outperforms ACE, and COSCO outperforms ACE as well for moderate SINR values. The effective SINR gain of PHORD over ACE at $K_s = 60 = 1.5N$ is roughly 3 dB. For the single steering vector case in Fig. 3.3, the performance of COSCO is very similar to that of PHORD, and both achieve higher probabilities of detection than ACE for the given parameter values.

Figures 3.4 and 3.5 show the cases when the target signal is range-spread and is distributed across $K_p = 8$ and $K_p = 2$ range cells, respectively. For the subspace signal case in Fig. 3.4, with $r = 9$, the one-step version of ACE outperforms its two-step version for $K_s = 60$, while for higher values of K_s the two versions give similar performances. On the other hand, both the one-step and two step versions of PHORD and COSCO give nearly identical detection probabilities and outperform ACE for the given scenarios. For the range spread ($K_p \geq 1$) and subspace detection ($r \geq 1$) scenarios considered, PHORD retains superior performance to ACE across the wide range of realistic secondary data sizes, $1.5N < Kp \leq 4N$.

Apart from their improved detection capabilities, one beneficial aspect of the PHORD and COSCO is that their two-step versions can be used with minimal performance loss compared to the one-step case; this is helpful since the one-step versions of all three detectors require the solution of an equation in the form of (3.11), which may be computationally prohibitive in some applied settings. In addition to the computational costs, the one-step version may also not always have a solution; for example, in Fig 3.5 (top), the relevant equation in the form of (3.11) does not have a solution for the PHORD and COSCO cases, [3] and hence only two-step versions of these detectors are used there. Throughout Fig. 3.5, where $K_p = 2$, little difference is observed between the one-step and two-step cases for all detectors. Furthermore, PHORD and COSCO exhibit improved detection performances compared to ACE.

Finally, in Fig. 3.6, we observe the performance of the three detectors under jamming. While the PHORD and COSCO are not specifically designed for jamming resistance (the orthogonal jamming term is only present in the H_0 hypothesis, unlike in AORD [65]), this beneficial trait can still be observed in the given empirical results. The jamming signal $\mathbf{j}$ in this case was constructed as a single ULA steering vector whose angle ω was chosen in each MC simulation as a uniform random variable in the range $[0.25, 0.35]$. Each element of the vector $\phi \in \mathbb{C}^{r \times K_p}$ was chosen as $\mathcal{CN}(0,1)$, and ϕ was scaled to achieve a desired jammer-to-noise ratio (JNR) defined as $\frac{1}{K_p}\phi^H \mathbf{j}^H \mathbf{R}^{-1}\mathbf{j}\phi$; the thresholds at each JNR value were set using 5×10^4 MC simulations to maintain the same Pfa (10^{-2}) for all the detectors. From Fig. 3.6, it can be seen that the one-step and two-step versions of PHORD and COSCO

---

[3] This is due to the fact that in the GLRT derivation for PHORD and COSCO, $n = \min(K_p, r) = 1$ for H_0, so a solution to (3.11) does not exist unless $\frac{NK_p}{K_p+K_s} < 1$.

give nearly identical performance and can withstand a much higher amount of jamming than ACE, though the one-step ACE becomes more robust to jamming in the range-spread case with an increase in K_p.

3.6   Conclusion

Two new detectors were proposed for the partially homogeneous scenario. Several connections were made between the newly proposed detectors and those established in the literature. The CFAR property of the proposed detectors was shown with resepct to both the structure and the scaling of the covariance matrix, and several geometric interpretations of their forms were given. An alternative derivation was also given for one of these detectors as a maxi-min SINR optimal matched filter. It was empirically shown that for the range of parameters considered, these detectors outperform the well-known ACE for low training data sizes; the reason for this benefit can be traced back to the use of a white-noise matched filter instead of the usual pseudowhitening operation used by ACE and other detectors. The proposed detectors also displayed a higher robustness in detection performance under the presence of jamming.

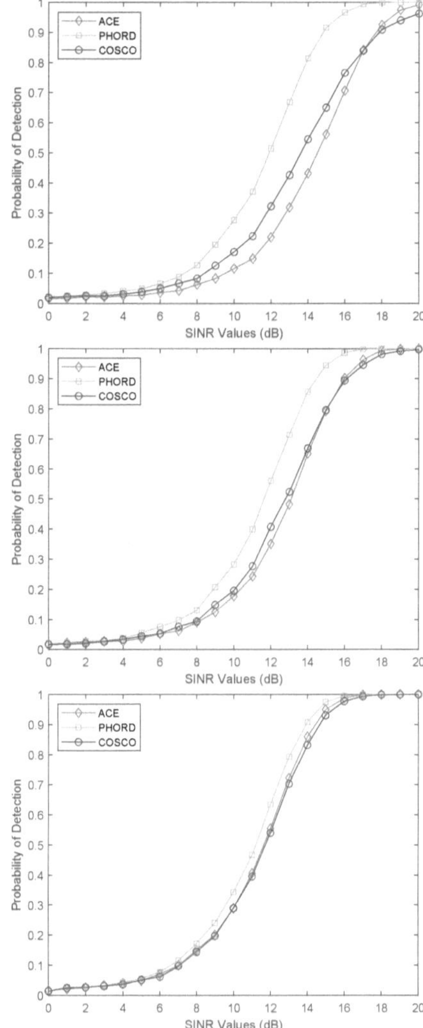

Figure 3.2: Probabilities of detection versus SINR of the ACE, COSCO, and PHORD, for $K_p = 1$, $r = 9$, $N = 40$, and, from top to bottom, $K_s = 60$, 80, and 160.

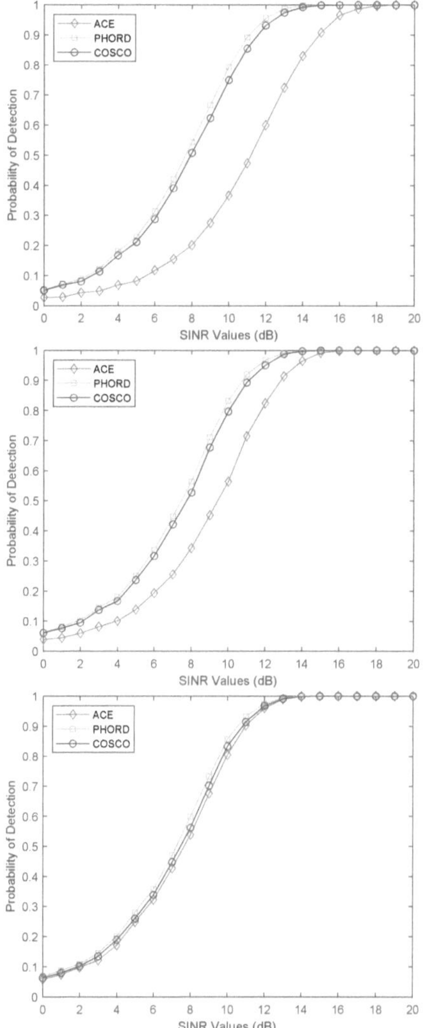

Figure 3.3: Probabilities of detection versus SINR of the ACE, COSCO, and PHORD, for $K_p = 1$, $r = 1$, $N = 40$, and, from top to bottom, $K_s = 60$, 80, and 160.

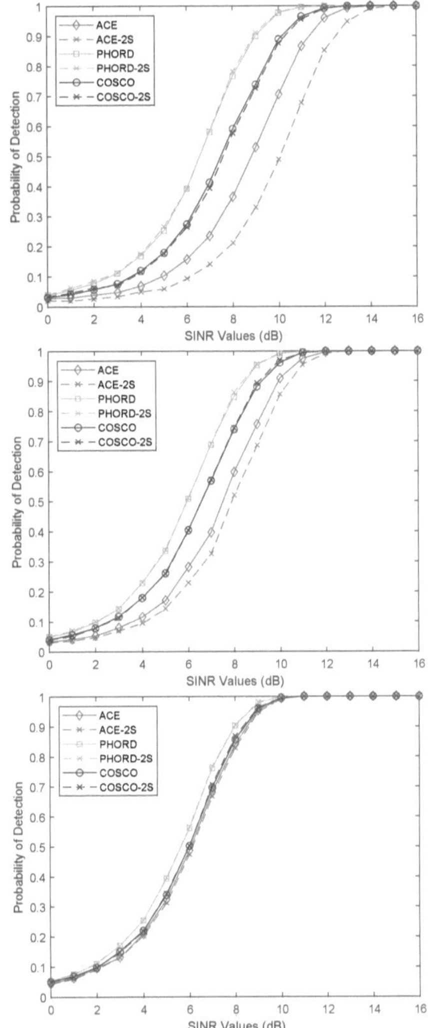

Figure 3.4: Probabilities of detection versus SINR of the ACE, COSCO, and PHORD, for $K_p = 8$, $r = 9$, $N = 40$, and, from top to bottom, $K_s = 60$, 80, and 160.

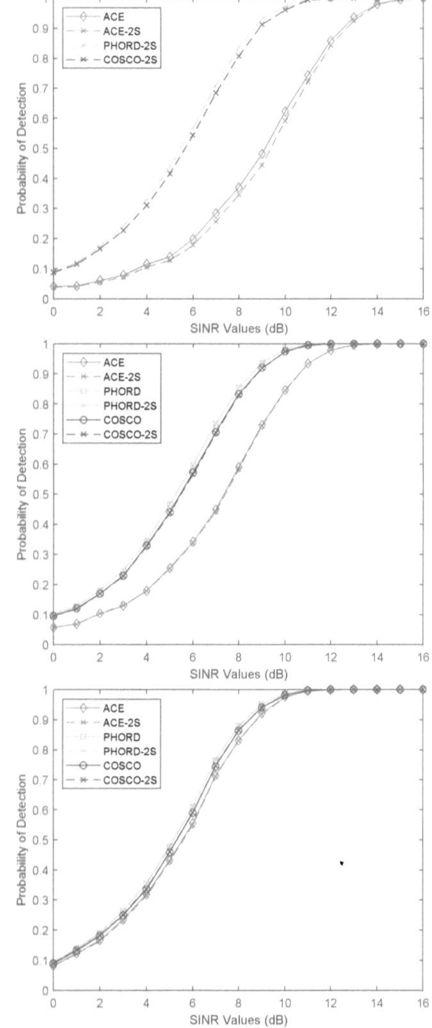

Figure 3.5: Probabilities of detection versus SINR of the ACE, COSCO, and PHORD, for $K_p = 2$, $r = 1$, $N = 40$, and, from top to bottom, $K_s = 60$, 80, and 160.

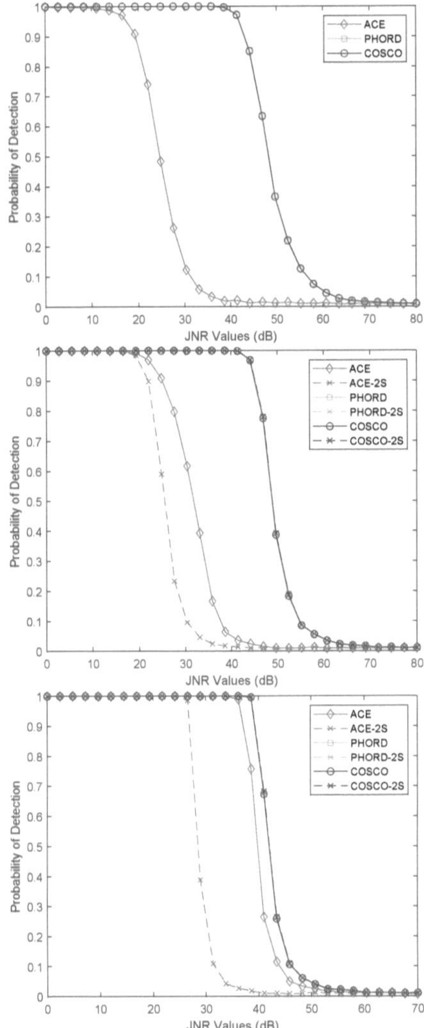

Figure 3.6: Probabilities of detection versus JNR of the ACE, COSCO, and PHORD, for $N = 40$, $r = 1$, and $K_p = 1$, $K_s = 80$, SINR = 16 dB (top), $K_p = 2$, $K_s = 80$, SINR = 16 dB (middle), and $K_p = 20$, $K_s = 800$, SINR = 8 dB (bottom).

Chapter 4: Subspace Signal Detection in Spectrally Symmetric Noise

4.1 Overview

In this chapter, we derive several detectors to test the presence or absence of a signal in noise which is assumed to be spectrally symmetric, meaning that its real and imaginary components are statistically independent. It is shown that when this assumption is incorporated into the detection problem, the resulting detectors, whether designed via a GLR, Rao, or Wald test approach, can give improved performance compared to the case when no spectral symmetry assumption is made.

4.2 Introduction

Adaptive detection of signals in Gaussian noise is a topic which has been studied for many decades in the signal processing community and continues to be an active area of research. In many cases, the underlying goal of this problem is to detect the presence or absence of a signal with a known signature and unknown scaling, contaminated by complex, circularly symmetric colored Gaussian noise whose covariance matrix is unknown [56]. A set of secondary data, assumed to be free of the signal in question, is taken to estimate the covariance matrix of the additive noise. Using both the primary and secondary data, one-step [56] and two-step (assuming the covariance matrix is known, then replacing it with an estimate) [80] approaches have been derived to find the Generalized Likelihood Ratio Test (GLRT), which is an alternative to the Neyman-Pearson-optimal likelihood ratio test (LRT) when certain parameters of the detection problem (in this case, the signal scaling and noise covariance) are unknown. However, the GLRT itself does not possess any optimality properties, and hence alternatives like the Rao and Wald tests [66, 55] have also been considered in the literature.

One of the ongoing challenges in adaptive detection is the scarcity of secondary data for estimating the covariance matrix of the colored Gaussian noise. In applications like adaptive radar [91, 47], where noise is often used to model both thermal noise present in the receiver as well as unwanted clutter returns, this scarcity is due to the fact that secondary data is often heterogeneous, and does not share the same statistical distribution as the noise present in the primary data. In this regard, it is desirable to derive detectors which achieve an acceptable performance using a limited amount of secondary data.

An important factor in the design of detectors for data-limited scenarios is incorporation of additional knowledge on the nature of the secondary data. One of the key characteristics of measured ground clutter data [16] which has received attention in the literature is that of spectral symmetry, meaning that the real and imaginary parts of such clutter returns can be approximated as being statistically independent. It can be shown that this property (which does not necessarily hold for arbitrary circular complex Gaussian data [38]) implies that the unknown covariance matrix of the colored noise has all real elements, which can greatly enhance its estimation. A number of papers [31, 49, 50, 34, 35] have considered the case of adaptive detection when the additive noise is known to be spectrally symmetric. However, all of these papers considered the case of point-like targets that are present in one range cell and belong to the subspace spanned by a single steering vector, and the GLRT and Wald detectors were only given either in a two-step or iterative manner. In practical applications, it is often of interest to detect a signal which can lie along a subspace spanned by several vectors [57, 67, 68, 19, 20], which is a modeling choice often made to account for possible uncertainties in the steering vector, for targets which may span multiple angle or Doppler values, or for detection across a certain swath of the Doppler/azimuth grid. In addition, range-spread modeling of the target signal is often useful for newer, higher-resolution radar systems, in which a potential target could be spread out over multiple range-cells which then need to be tested for target presence simultaneously [24]. Moreover, since the two-step versions of the GLR and Wald tests do not possess any optimality properties, their (approximate) one-step versions are also of interest as additional alternatives to consider.

Toward this end, in this chapter we derive and investigate several detectors using the approximate GLR, Rao, and approximate Wald tests for the case of subspace and/or range-spread signals which incorporate knowledge of the noise spectral symmetry. These detectors are given for both the classical fully homogeneous data case, as considered in [56, 80], and for the partially homogeneous case [60], wherein the latter assumes an unknown, positive scaling between the covariance matrices of the primary and secondary data. Numerical simulations are provided which compare the performances of these detectors and show the improvement obtained through the incorporation of the spectral symmetry constraint to subspace and range-spread signal detection.

4.3 Design of Detectors

4.3.1 Problem Setup

The detection problem we are interested in can be described as a hypothesis test of the form

$$\begin{cases} H_1 : \mathbf{X} = \mathbf{V}\mathbf{\Theta} + \mathbf{N}, & \mathbf{x}_k = \mathbf{n}_k, \ k = 1, 2, \ldots, K_s \\ H_0 : \mathbf{X} = \mathbf{N}, & \mathbf{x}_k = \mathbf{n}_k, \ k = 1, 2, \ldots, K_s, \end{cases} \tag{4.1}$$

where $\mathbf{V} \in \mathbb{C}^{N \times r}$, $1 \leq r < N$, is a known matrix whose orthonormal columns span the signal subspace, and $\mathbf{\Theta} \in \mathbb{C}^{r \times K_p}$ represents the deterministic, unknown coordinates of the hypothetical signal within this subspace. Note that if $\mathbf{V}$ didn't have orthonormal columns, we could simply take its QR decomposition as $\mathbf{V} = \mathbf{Q}\mathbf{R}_Q$ and redefine the $\mathbf{V}$ and $\mathbf{\Theta}$ matrices as $\mathbf{V}_{new} = \mathbf{Q} \in \mathbb{C}^{N \times r}$ (which would now have orthonormal columns) and $\mathbf{\Theta}_{new} = \mathbf{R}_Q\mathbf{\Theta}$.

The matrix $\mathbf{X}$ represents the Cell(s) Under Test (CUT), while $\mathbf{x}_k \in \mathbb{C}^{N \times 1}$ are the secondary signal-free training data. For the partially homogeneous environment, the additive noise is modeled such that each column of $\mathbf{N}$ is independent and identically distributed (iid) as $\mathcal{CN}(\mathbf{0}, \gamma\mathbf{R})$ and $\mathbf{n}_k \sim \mathcal{CN}(\mathbf{0}, \mathbf{R})$, where $\gamma > 0$ is an unknown quantity, while for the homogeneous environment we simply let $\gamma = 1$.

4.3.2 Approximate GLRT for the Partially Homogeneous Environment

Letting $\mathbf{X}_s = [\mathbf{x}_1 \ldots \mathbf{x}_{K_s}]$ be the collection of secondary training data, the expression to be evaluated for the one-step GLRT is given as

$$\frac{\max\limits_{\Theta} \max\limits_{\gamma} \max\limits_{\mathbf{R}} f_{H_1}(\mathbf{X}, \mathbf{X}_s; \mathbf{R}, \gamma, \Theta)}{\max\limits_{\gamma} \max\limits_{\mathbf{R}} f_{H_0}(\mathbf{X}, \mathbf{X}_s; \mathbf{R}, \gamma)} \mathop{\gtrless}\limits_{H_0}^{H_1} \xi. \tag{4.2}$$

Then, the likelihood function for H_1 can be written as

$$f_{H_1}(\mathbf{X}, \mathbf{X}_s; \mathbf{R}, \gamma, \Theta) = \frac{1}{\gamma^{NK_p} \det(\mathbf{R})^{K_p+K_s}} \times$$
$$\exp(-\operatorname{tr}\{\frac{1}{\gamma}(\mathbf{X} - \mathbf{V}\Theta)^H \mathbf{R}^{-1}(\mathbf{X} - \mathbf{V}\Theta) + \mathbf{X}_s^H \mathbf{R}^{-1}\mathbf{X}_s\}). \tag{4.3}$$

For the case of spectrally symmetric noise, we can rewrite (4.3) as follows. Let us define the "stacked" versions of $\mathbf{V}$, $\mathbf{X}$, $\mathbf{N}$, and $\mathbf{R}$ as

$$\overline{\mathbf{V}} = \begin{bmatrix} \mathbf{V}_{Re} & -\mathbf{V}_{Im} \\ \mathbf{V}_{Im} & \mathbf{V}_{Re} \end{bmatrix}, \tag{4.4}$$

$$\overline{\mathbf{X}} = \begin{bmatrix} \mathbf{X}_{Re} \\ \mathbf{X}_{Im} \end{bmatrix}, \tag{4.5}$$

$$\overline{\mathbf{N}} = \begin{bmatrix} \mathbf{N}_{Re} \\ \mathbf{N}_{Im} \end{bmatrix}, \tag{4.6}$$

and

$$\overline{\mathbf{R}} = \begin{bmatrix} \mathbf{R} & \mathbf{0} \\ \mathbf{0} & \mathbf{R} \end{bmatrix}, \tag{4.7}$$

where $\mathbf{V}_{Re}$ and $\mathbf{V}_{Im}$ denote the real and imaginary parts of $\mathbf{V}$, and the same with the other variables (note that $\mathbf{R}_{Im} = \mathbf{0}$). Furthermore, defining $\mathbf{Y}_{Re} = (\mathbf{X} - \mathbf{V}\Theta)_{Re}$ and $\mathbf{Y}_{Im} = (\mathbf{X} - \mathbf{V}\Theta)_{Im}$, we can likewise stack $\mathbf{Y}_{Re}$ and $\mathbf{Y}_{Im}$ together into $\overline{\mathbf{Y}}$ as was done for $\overline{\mathbf{X}}$ and $\overline{\mathbf{N}}$. In particular, we have

$$\overline{\mathbf{Y}} = \overline{\mathbf{X}} - \overline{\mathbf{V}}\,\overline{\Theta}, \tag{4.8}$$

66

where $\overline{\Theta}$ is likewise stacked from Θ_{Re} and Θ_{Im} in the same manner as $\overline{\mathbf{X}}$ and $\overline{\mathbf{N}}$. Finally, we can stack the real and imaginary components of $\mathbf{X}_s$ into $\overline{\mathbf{X}}_s$ in a similar fashion. It can readily be ascertained that in the presence of spectral symmetry, both $\overline{\mathbf{N}}$ and $\overline{\mathbf{X}}_s$ are distributed as $\mathcal{N}(0, \frac{1}{2}\overline{\mathbf{R}})$. Thus, we can alternatively write the H_1 likelihood in the spectrally symmetric case as

$$f_{H_1}(\mathbf{X}, \mathbf{X}_s; \mathbf{R}, \gamma, \Theta) = \frac{2^{2NK_p}}{\gamma^{2NK_p} \det{(\overline{\mathbf{R}})}^{K_p + K_s}} \exp(-\frac{1}{2}\operatorname{tr}\{\frac{1}{\gamma}\overline{\mathbf{Y}}^T\overline{\mathbf{R}}^{-1}\overline{\mathbf{Y}} + \overline{\mathbf{X}}_s^T\overline{\mathbf{R}}^{-1}\overline{\mathbf{X}}_s\})$$

$$(4.9)$$

$$\propto \frac{1}{\gamma^{2NK_p} \det{(\mathbf{R})}^{2(K_p + K_s)}} \exp(-\frac{1}{2}\operatorname{tr}\{\frac{1}{\gamma}(\mathbf{Y}_{Re}^T\mathbf{R}^{-1}\mathbf{Y}_{Re} + \mathbf{Y}_{Im}^T\mathbf{R}^{-1}\mathbf{Y}_{Im})\})\times$$

$$\exp(-\frac{1}{2}\operatorname{tr}\{\mathbf{X}_{s,Re}^T\mathbf{R}^{-1}\mathbf{X}_{s,Re} + \mathbf{X}_{s,Im}^T\mathbf{R}^{-1}\mathbf{X}_{s,Im}\})$$

$$= \frac{1}{\gamma^{2NK_p} \det{(\mathbf{R})}^{2(K_p + K_s)}} \exp(-\frac{1}{2}\operatorname{tr}\{\mathbf{R}^{-1}\frac{1}{\gamma}(\mathbf{Y}_{Re}\mathbf{Y}_{Re}^T + \mathbf{Y}_{Im}\mathbf{Y}_{Im}^T)\})\times$$

$$\exp(-\frac{1}{2}\operatorname{tr}\{\mathbf{R}^{-1}(\mathbf{X}_{s,Re}\mathbf{X}_{s,Re}^T + \mathbf{X}_{s,Im}\mathbf{X}_{s,Im}^T)\}).$$

The purpose of rewriting the likelihood function in this manner is to incorporate the prior knowledge that the real and imaginary components of the received signal noise are statistically independent. From the last expression in (4.9), it can readily be found that the maximum likelihood (ML) estimate of $\mathbf{R}$ under H_1 is given as

$$\widehat{\mathbf{R}}_1 = \frac{1}{2(K_p + K_s)}[\frac{1}{\gamma}(\mathbf{Y}_{Re}\mathbf{Y}_{Re}^T + \mathbf{Y}_{Im}\mathbf{Y}_{Im}^T) + \mathbf{X}_{s,Re}\mathbf{X}_{s,Re}^T + \mathbf{X}_{s,Im}\mathbf{X}_{s,Im}^T]. \qquad (4.10)$$

Putting this quantity back into the likelihood and ignoring unrelated scaling constants, we get

$$f_{H_1}(\mathbf{X}, \mathbf{X}_s; \widehat{\mathbf{R}}_1, \gamma, \Theta) \propto \frac{1}{\gamma^{2NK_p}} \times$$

$$\frac{1}{\det{(\frac{1}{\gamma}(\mathbf{Y}_{Re}\mathbf{Y}_{Re}^T + \mathbf{Y}_{Im}\mathbf{Y}_{Im}^T) + \mathbf{X}_{s,Re}\mathbf{X}_{s,Re}^T + \mathbf{X}_{s,Im}\mathbf{X}_{s,Im}^T)}^{2(K_p + K_s)}}$$

From the above formulation, it may not readily appear that the optimization of the likelihood over Θ has any closed-form solution, and indeed the authors of [31] resorted to a sophisticated iterative optimization approach for the special case of $K_p = r = \gamma = 1$. Here, we present an approximate one-step GLRT and then demonstrate the efficacy of the resulting detector. First, note that we can rewrite the determinant term in the likelihood (ignoring the exponent for the time being) as

$$\det\left(\mathbf{I} + \frac{1}{\gamma}[\mathbf{S}_{ss}^{-\frac{1}{2}}\mathbf{Y}_{Re}\mathbf{Y}_{Re}^T\mathbf{S}_{ss}^{-\frac{1}{2}} + \mathbf{S}_{ss}^{-\frac{1}{2}}\mathbf{Y}_{Im}\mathbf{Y}_{Im}^T\mathbf{S}_{ss}^{-\frac{1}{2}}]\right) \tag{4.11}$$

where we let $\mathbf{S}_{ss} = \mathbf{X}_{s,Re}\mathbf{X}_{s,Re}^T + \mathbf{X}_{s,Im}\mathbf{X}_{s,Im}^T$. Note that the matrix

$$\left(\mathbf{S}_{ss}^{-\frac{1}{2}}\mathbf{Y}_{Re}\mathbf{Y}_{Re}^T\mathbf{S}_{ss}^{-\frac{1}{2}} + \mathbf{S}_{ss}^{-\frac{1}{2}}\mathbf{Y}_{Im}\mathbf{Y}_{Im}^T\mathbf{S}_{ss}^{-\frac{1}{2}}\right)$$

is positive semidefinite, and hence all of its eigenvalues λ_i are greater than or equal to zero. Thus, with the linear signal model, we adopt a heuristic based on the following approximation: we can write

$$\begin{aligned}
&\arg\min_{\boldsymbol{\Theta}_{Re},\boldsymbol{\Theta}_{Im}} \det\left(\mathbf{I} + \frac{1}{\gamma}[\mathbf{S}_{ss}^{-\frac{1}{2}}\mathbf{Y}_{Re}\mathbf{Y}_{Re}^T\mathbf{S}_{ss}^{-\frac{1}{2}} + \mathbf{S}_{ss}^{-\frac{1}{2}}\mathbf{Y}_{Im}\mathbf{Y}_{Im}^T\mathbf{S}_{ss}^{-\frac{1}{2}}]\right) \\
&\approx \arg\min_{\boldsymbol{\Theta}_{Re},\boldsymbol{\Theta}_{Im}} \operatorname{tr}\left(\mathbf{I} + \frac{1}{\gamma}[\mathbf{S}_{ss}^{-\frac{1}{2}}\mathbf{Y}_{Re}\mathbf{Y}_{Re}^T\mathbf{S}_{ss}^{-\frac{1}{2}} + \mathbf{S}_{ss}^{-\frac{1}{2}}\mathbf{Y}_{Im}\mathbf{Y}_{Im}^T\mathbf{S}_{ss}^{-\frac{1}{2}}]\right) \\
&= \arg\min_{\boldsymbol{\Theta}} \operatorname{tr}\left(\mathbf{I} + \frac{1}{\gamma}[\overline{\mathbf{S}}_{ss}^{-\frac{1}{2}}\overline{\mathbf{Y}\mathbf{Y}}^T\overline{\mathbf{S}}_{ss}^{-\frac{1}{2}}]\right) \\
&= \arg\min_{\boldsymbol{\Theta}} \operatorname{tr}\left(\overline{\mathbf{S}}_{ss}^{-\frac{1}{2}}\overline{\mathbf{Y}\mathbf{Y}}^T\overline{\mathbf{S}}_{ss}^{-\frac{1}{2}}\right) \\
&= \arg\min_{\boldsymbol{\Theta}} \|\overline{\mathbf{S}}_{ss}^{-\frac{1}{2}}(\overline{\mathbf{X}} - \overline{\mathbf{V}}\,\overline{\boldsymbol{\Theta}})\|_F^2, \tag{4.12}
\end{aligned}$$

where we let

$$\overline{\mathbf{S}}_{ss} = \begin{bmatrix} \mathbf{S}_{ss} & 0 \\ 0 & \mathbf{S}_{ss} \end{bmatrix}. \tag{4.13}$$

Because the last line of (4.12) is a simple weighted least-squares problem, the estimate of $\overline{\boldsymbol{\Theta}}$ can then be found as

$$\widehat{\overline{\boldsymbol{\Theta}}} = \begin{bmatrix} \widehat{\boldsymbol{\Theta}}_{Re} \\ \widehat{\boldsymbol{\Theta}}_{Im} \end{bmatrix} = (\overline{\mathbf{V}}^T\overline{\mathbf{S}}_{ss}^{-1}\overline{\mathbf{V}})^{-1}\overline{\mathbf{V}}^T\overline{\mathbf{S}}_{ss}^{-1}\overline{\mathbf{X}}. \tag{4.14}$$

Putting the estimates $\widehat{\boldsymbol{\Theta}}_{Re}$ and $\widehat{\boldsymbol{\Theta}}_{Im}$ into $\mathbf{Y}_{Re}$ and $\mathbf{Y}_{Im}$, we get

$$\widehat{\mathbf{Y}}_{Re} = \widehat{\mathbf{X}}_{Re} - (\mathbf{V}_{Re}\widehat{\boldsymbol{\Theta}}_{Re} - \mathbf{V}_{Im}\widehat{\boldsymbol{\Theta}}_{Im}) \tag{4.15}$$

and

$$\widehat{\mathbf{Y}}_{Im} = \widehat{\mathbf{X}}_{Im} - (\mathbf{V}_{Re}\widehat{\boldsymbol{\Theta}}_{Im} + \mathbf{V}_{Im}\widehat{\boldsymbol{\Theta}}_{Re}). \tag{4.16}$$

Using our newly-found values of $\widehat{\mathbf{Y}}_{Re}$ and $\widehat{\mathbf{Y}}_{Im}$, we can write the H_1 likelihood function as

$$f_{H_1}(\mathbf{X}, \mathbf{X}_s; \widehat{\mathbf{R}}_1, \gamma, \widehat{\boldsymbol{\Theta}}) \propto \frac{1}{\gamma^{2NK_p}} \times$$

$$\frac{1}{\det\left(\mathbf{I} + \frac{1}{\gamma}[\mathbf{S}_{ss}^{-\frac{1}{2}}\mathbf{Y}_{Re}\mathbf{Y}_{Re}^{T}\mathbf{S}_{ss}^{-\frac{1}{2}} + \mathbf{S}_{ss}^{-\frac{1}{2}}\mathbf{Y}_{Im}\mathbf{Y}_{Im}^{T}\mathbf{S}_{ss}^{-\frac{1}{2}}]\right)^{2(K_p+K_s)}} \tag{4.17}$$

Now, to do this, note that we can write the $2(K_p + K_s)$th root of the denominator in (4.17) as

$$\gamma^{\frac{NK_p}{K_p+K_s}} \prod_{i=1}^{n}(1 + \frac{\lambda_i}{\gamma}), \tag{4.18}$$

where n is the number of nonzero eigenvalues λ_i of the matrix $\mathbf{S}_{ss}^{-\frac{1}{2}}\mathbf{Y}_{Re}\mathbf{Y}_{Re}^{T}\mathbf{S}_{ss}^{-\frac{1}{2}} + \mathbf{S}_{ss}^{-\frac{1}{2}}\mathbf{Y}_{Im}\mathbf{Y}_{Im}^{T}\mathbf{S}_{ss}^{-\frac{1}{2}}$. Taking the logarithm of this quantity, and subsequently setting the derivative with respect to γ to zero, yields the expression

$$\frac{NK_p}{K_p + K_s} - \sum_{i=1}^{n}\frac{\lambda_i}{\lambda_i + \gamma} = 0. \tag{4.19}$$

Clearly, when $\frac{NK_p}{K_p+K_s} < n$, there exists a unique, positive value of $\gamma = \widehat{\gamma}_1$ which satisfies (4.19) and hence maximizes the value in (4.17), since the summation term in (4.19) is a monotonic function of $\gamma > 0$ with range $[0, n)$. Turning now to the H_0

case, we can write the likelihood function as

$$f_{H_0}(\mathbf{X}, \mathbf{X}_s; \mathbf{R}, \gamma) = \frac{2^{2NK_p}}{\gamma^{2NK_p} \det{(\overline{\mathbf{R}})}^{K_p+K_s}} \exp(-\frac{1}{2}\operatorname{tr}\{\frac{1}{\gamma}\overline{\mathbf{X}}^T\overline{\mathbf{R}}^{-1}\overline{\mathbf{X}} + \overline{\mathbf{X}}_s^T\overline{\mathbf{R}}^{-1}\overline{\mathbf{X}}_s\})$$

$$\propto \frac{1}{\gamma^{2NK_p} \det{(\mathbf{R})}^{2(K_p+K_s)}} \exp(-\frac{1}{2}\operatorname{tr}\{\mathbf{R}^{-1}\frac{1}{\gamma}(\mathbf{X}_{Re}\mathbf{X}_{Re}^T + \mathbf{X}_{Im}\mathbf{X}_{Im}^T)\})\times$$

$$\exp(-\frac{1}{2}\operatorname{tr}\{\mathbf{R}^{-1}(\mathbf{X}_{s,Re}\mathbf{X}_{s,Re}^T + \mathbf{X}_{s,Im}\mathbf{X}_{s,Im}^T)\}). \quad (4.20)$$

From (4.20), it can be shown that the ML estimate of $\mathbf{R}$ under H_0 is given as

$$\widehat{\mathbf{R}}_0 = \frac{1}{2(K_p + K_s)}\left[\frac{1}{\gamma}(\mathbf{X}_{Re}\mathbf{X}_{Re}^T + \mathbf{X}_{Im}\mathbf{X}_{Im}^T) + \mathbf{X}_{s,Re}\mathbf{X}_{s,Re}^T + \mathbf{X}_{s,Im}\mathbf{X}_{s,Im}^T\right]. \quad (4.21)$$

Putting this estimate back into (4.20), we can rewrite the likelihood function as

$$f_{H_0}(\mathbf{X}, \mathbf{X}_s; \widehat{\mathbf{R}}_0, \gamma) \propto \frac{1}{\gamma^{2NK_p}} \times$$

$$\frac{1}{\det{(\mathbf{I} + \frac{1}{\gamma}[\mathbf{S}_{ss}^{-\frac{1}{2}}\mathbf{X}_{Re}\mathbf{X}_{Re}^T\mathbf{S}_{ss}^{-\frac{1}{2}} + \mathbf{S}_{ss}^{-\frac{1}{2}}\mathbf{X}_{Im}\mathbf{X}_{Im}^T\mathbf{S}_{ss}^{-\frac{1}{2}}])}^{2(K_p+K_s)}}. \quad (4.22)$$

Similarly to the H_1 case, the optimal $\gamma = \widehat{\gamma}_0$ for H_0 which maximizes (4.22) can be found as the solution of (4.19) with the λ_i now corresponding to the eigenvalues of the matrix $\mathbf{S}_{ss}^{-\frac{1}{2}}\mathbf{X}_{Re}\mathbf{X}_{Re}^T\mathbf{S}_{ss}^{-\frac{1}{2}} + \mathbf{S}_{ss}^{-\frac{1}{2}}\mathbf{X}_{Im}\mathbf{X}_{Im}^T\mathbf{S}_{ss}^{-\frac{1}{2}}$. Finally, we can give the approximate one-step GLRT statistic for the spectrally symmetric, partially homogeneous environment as

$$\frac{f_{H_1}(\mathbf{X}, \mathbf{X}_s; \widehat{\mathbf{R}}_1, \widehat{\gamma}_1, \widehat{\mathbf{\Theta}}_{H_1})}{f_{H_0}(\mathbf{X}, \mathbf{X}_s; \widehat{\mathbf{R}}_0, \widehat{\gamma}_0)} \propto \left(\frac{\widehat{\gamma}_0}{\widehat{\gamma}_1}\right)^{\frac{NK_p}{K_p+K_s}} \times$$

$$\frac{\det{(\mathbf{I} + \frac{1}{\widehat{\gamma}_0}[\mathbf{S}_{ss}^{-\frac{1}{2}}\mathbf{X}_{Re}\mathbf{X}_{Re}^T\mathbf{S}_{ss}^{-\frac{1}{2}} + \mathbf{S}_{ss}^{-\frac{1}{2}}\mathbf{X}_{Im}\mathbf{X}_{Im}^T\mathbf{S}_{ss}^{-\frac{1}{2}}])}{\det{(\mathbf{I} + \frac{1}{\widehat{\gamma}_1}[\mathbf{S}_{ss}^{-\frac{1}{2}}\widehat{\mathbf{Y}}_{Re}\widehat{\mathbf{Y}}_{Re}^T\mathbf{S}_{ss}^{-\frac{1}{2}} + \mathbf{S}_{ss}^{-\frac{1}{2}}\widehat{\mathbf{Y}}_{Im}\widehat{\mathbf{Y}}_{Im}^T\mathbf{S}_{ss}^{-\frac{1}{2}}])}, \quad (4.23)$$

where we took the $2(K_p + K_s)$th root of the final detection statistic.

4.3.3 Approximate GLRT for the Fully Homogeneous Environment

For the case of full homogeneity, we have $\gamma = 1$, so that the relevant form for the GLRT now becomes

$$\frac{\max_{\Theta} \max_{\mathbf{R}} f_{H_1}(\mathbf{X}, \mathbf{X}_s; \mathbf{R}, \Theta)}{\max_{\mathbf{R}} f_{H_0}(\mathbf{X}, \mathbf{X}_s; \mathbf{R})} \underset{H_0}{\overset{H_1}{\gtrless}} \xi. \tag{4.24}$$

The likelihood functions $f_{H_1}(\mathbf{X}, \mathbf{X}_s; \mathbf{R}, \Theta)$ and $f_{H_0}(\mathbf{X}, \mathbf{X}_s; \mathbf{R})$ are now the same as those in (4.9) and (4.20) but with $\gamma = 1$. Similarly, the estimates $\widehat{\mathbf{R}}_1$ and $\widehat{\mathbf{R}}_0$ are the same as those in (4.10) and (4.21) but with $\gamma = 1$, and with $\mathbf{Y}_{Re}$ and $\mathbf{Y}_{Im}$ replaced with their ML estimates $\widehat{\mathbf{Y}}_{Re}$ and $\widehat{\mathbf{Y}}_{Im}$ in (4.15) and (4.16). The approximate one-step GLRT statistic for the homogeneous environment can then be given as

$$\frac{f_{H_1}(\mathbf{X}, \mathbf{X}_s; \widehat{\mathbf{R}}_1, \widehat{\Theta}_1)}{f_{H_0}(\mathbf{X}, \mathbf{X}_s; \widehat{\mathbf{R}}_0)} \propto \frac{\det\left(\mathbf{I} + \mathbf{S}_{ss}^{-\frac{1}{2}}\mathbf{X}_{Re}\mathbf{X}_{Re}^T\mathbf{S}_{ss}^{-\frac{1}{2}} + \mathbf{S}_{ss}^{-\frac{1}{2}}\mathbf{X}_{Im}\mathbf{X}_{Im}^T\mathbf{S}_{ss}^{-\frac{1}{2}}\right)}{\det\left(\mathbf{I} + \mathbf{S}_{ss}^{-\frac{1}{2}}\widehat{\mathbf{Y}}_{Re}\widehat{\mathbf{Y}}_{Re}^T\mathbf{S}_{ss}^{-\frac{1}{2}} + \mathbf{S}_{ss}^{-\frac{1}{2}}\widehat{\mathbf{Y}}_{Im}\widehat{\mathbf{Y}}_{Im}^T\mathbf{S}_{ss}^{-\frac{1}{2}}\right)}. \tag{4.25}$$

4.3.4 Two-step GLRT for the Partially Homogeneous Environment

To obtain the two-step GLRT, we first assume the true covariance matrix $\mathbf{R}$ and derive the statistic

$$\frac{\max_{\Theta} \max_{\gamma} f_{H_1}(\mathbf{X}, \mathbf{X}_s; \gamma, \Theta)}{\max_{\gamma} f_{H_0}(\mathbf{X}, \mathbf{X}_s; \gamma)} \underset{H_0}{\overset{H_1}{\gtrless}} \xi. \tag{4.26}$$

as a function of $\mathbf{R}$. We then replace $\mathbf{R}$ by its ML estimate using secondary data. To proceed, we can write the H_1 likelihood function, ignoring any unrelated scalings, as

$$f_{H_1}(\mathbf{X}, \mathbf{X}_s; \gamma, \Theta) \propto \frac{1}{\gamma^{2NK_p}} \exp\left(-\frac{1}{2}\operatorname{tr}\{\frac{1}{\gamma}\overline{\mathbf{Y}}^T\mathbf{R}^{-1}\overline{\mathbf{Y}}\}\right) \tag{4.27}$$

for the spectrally symmetric case. Taking the derivative of the above with respect to γ and setting it equal to zero gives its optimal estimate for H_1 as $\widehat{\gamma}_1 =$

$\frac{1}{4NK_p}\,\text{tr}(\overline{\mathbf{Y}}^T\overline{\mathbf{R}}^{-1}\overline{\mathbf{Y}})$. Putting this value back in (4.27) and ignoring any scaling constants gives

$$f_{H_1}(\mathbf{X},\mathbf{X}_s;\widehat{\gamma}_1,\boldsymbol{\Theta}) \propto \frac{1}{\text{tr}(\overline{\mathbf{Y}}^T\overline{\mathbf{R}}^{-1}\overline{\mathbf{Y}})^{2NK_p}} \tag{4.28}$$

Noting that $\text{tr}(\overline{\mathbf{Y}}^T\overline{\mathbf{R}}^{-1}\overline{\mathbf{Y}}) = \|\overline{\mathbf{R}}^{-\frac{1}{2}}(\overline{\mathbf{X}} - \overline{\mathbf{V}}\,\boldsymbol{\Theta})\|_F^2$, we can find the weighted least-squares (and ML) optimal estimate of $\overline{\boldsymbol{\Theta}}$ as $\widehat{\overline{\boldsymbol{\Theta}}} = (\overline{\mathbf{V}}^T\overline{\mathbf{R}}^{-1}\mathbf{V})^{-1}\overline{\mathbf{V}}^T\overline{\mathbf{R}}^{-1}\overline{\mathbf{X}}$. Putting this back into (4.28) and simplifying, we get

$$f_{H_1}(\mathbf{X},\mathbf{X}_s;\widehat{\gamma}_1,\widehat{\boldsymbol{\Theta}}) \propto \frac{1}{[\text{tr}(\overline{\mathbf{X}}^T\overline{\mathbf{R}}^{-1}\overline{\mathbf{X}}) - \text{tr}(\overline{\mathbf{X}}^T\overline{\mathbf{R}}^{-1}\overline{\mathbf{V}}(\overline{\mathbf{V}}^T\overline{\mathbf{R}}^{-1}\overline{\mathbf{V}})^{-1}\overline{\mathbf{V}}^T\overline{\mathbf{R}}^{-1}\overline{\mathbf{X}})]^{2NK_p}} \tag{4.29}$$

Following an analogous procedure for the H_0 likelihood, we can find its maximized version as

$$f_{H_0}(\mathbf{X},\mathbf{X}_s;\widehat{\gamma}_0) \propto \frac{1}{[\text{tr}(\overline{\mathbf{X}}^T\overline{\mathbf{R}}^{-1}\overline{\mathbf{X}})]^{2NK_p}} \tag{4.30}$$

Forming a ratio of the two optimized likelihoods, taking their $2NK_p$th root, and dividing the numerator and denominator by $\text{tr}(\overline{\mathbf{X}}^T\overline{\mathbf{R}}^{-1}\overline{\mathbf{X}})$, we can find the GLRT statistic assuming a known $\overline{\mathbf{R}}$ to be proportional to

$$\frac{\text{tr}(\overline{\mathbf{X}}^T\overline{\mathbf{R}}^{-1}\overline{\mathbf{V}}(\overline{\mathbf{V}}^T\overline{\mathbf{R}}^{-1}\overline{\mathbf{V}})^{-1}\overline{\mathbf{V}}^T\overline{\mathbf{R}}^{-1}\overline{\mathbf{X}})}{\text{tr}(\overline{\mathbf{X}}^T\overline{\mathbf{R}}^{-1}\overline{\mathbf{X}})}. \tag{4.31}$$

Replacing the unknown $\overline{\mathbf{R}}$ with its estimate in (4.13) computed from the secondary data, we get the two-step GLRT statistic for the partially homogeneous environment given by

$$\frac{\text{tr}(\overline{\mathbf{X}}^T\overline{\mathbf{S}}_{ss}^{-1}\overline{\mathbf{V}}(\overline{\mathbf{V}}^T\overline{\mathbf{S}}_{ss}^{-1}\overline{\mathbf{V}})^{-1}\overline{\mathbf{V}}^T\overline{\mathbf{S}}_{ss}^{-1}\overline{\mathbf{X}})}{\text{tr}(\overline{\mathbf{X}}^T\overline{\mathbf{S}}_{ss}^{-1}\overline{\mathbf{X}})}. \tag{4.32}$$

4.3.5 Two-step GLRT for the Fully Homogeneous Environment

When $\gamma = 1$ for the case of the fully homogeneous environment, and when $\overline{\mathbf{R}}$ is initially assumed to be known for the two-step GLRT, then the only parameter left to optimize over is $\overline{\boldsymbol{\Theta}}$ for H_1. This estimate can readily be found from (4.9), after setting $\gamma = 1$ and ignoring unrelated terms, as

$$\widehat{\overline{\boldsymbol{\Theta}}} = (\overline{\mathbf{V}}^T \overline{\mathbf{R}}^{-1} \overline{\mathbf{V}})^{-1} \overline{\mathbf{V}}^T \overline{\mathbf{R}}^{-1} \overline{\mathbf{X}}. \tag{4.33}$$

Using this value along with (4.9) and (4.30), and ignoring any unrelated scaling constants, we can see that the ratio of the optimized likelihoods can be written as

$$\frac{f_{H_1}(\mathbf{X}, \mathbf{X}_s; \widehat{\boldsymbol{\Theta}}_1)}{f_{H_0}(\mathbf{X}, \mathbf{X}_s)} \propto \frac{\exp(-\frac{1}{2}\operatorname{tr}((\overline{\mathbf{X}} - \overline{\mathbf{V}}\,\widehat{\overline{\boldsymbol{\Theta}}})^T \overline{\mathbf{R}}^{-1}(\overline{\mathbf{X}} - \overline{\mathbf{V}}\,\widehat{\overline{\boldsymbol{\Theta}}}))}{\exp(-\frac{1}{2}\operatorname{tr}(\overline{\mathbf{X}}\,\overline{\mathbf{R}}^{-1}\overline{\mathbf{X}}))}, \tag{4.34}$$

and it can also readily be observed that

$$\ln\left(\frac{f_{H_1}(\mathbf{X}, \mathbf{X}_s; \widehat{\boldsymbol{\Theta}}_1)}{f_{H_0}(\mathbf{X}, \mathbf{X}_s)}\right) \propto \operatorname{tr}(\overline{\mathbf{X}}^T \overline{\mathbf{R}}^{-1} \overline{\mathbf{V}} (\overline{\mathbf{V}}^T \overline{\mathbf{R}}^{-1} \overline{\mathbf{V}})^{-1} \overline{\mathbf{V}}^T \overline{\mathbf{R}}^{-1} \overline{\mathbf{X}}). \tag{4.35}$$

Replacing the unknown $\overline{\mathbf{R}}$ with its estimate in (4.13) computed from the secondary data, we get the two-step GLRT statistic for the homogeneous environment given by

$$\operatorname{tr}(\overline{\mathbf{X}}^T \overline{\mathbf{S}}_{ss}^{-1} \overline{\mathbf{V}} (\overline{\mathbf{V}}^T \overline{\mathbf{S}}_{ss}^{-1} \overline{\mathbf{V}})^{-1} \overline{\mathbf{V}}^T \overline{\mathbf{S}}_{ss}^{-1} \overline{\mathbf{X}}). \tag{4.36}$$

4.3.6 Rao Test for the Partially Homogeneous Environment

The Rao test is given by the expression [55]

$$\left.\frac{\partial \ln f(\mathbf{x}; \boldsymbol{\theta})}{\partial \boldsymbol{\theta}_r}\right|^T_{\boldsymbol{\theta} = \widetilde{\boldsymbol{\theta}}} [\mathbf{I}_{\mathrm{F}}^{-1}(\widetilde{\boldsymbol{\theta}})]_{1:r,1:r} \left.\frac{\partial \ln f(\mathbf{x}; \boldsymbol{\theta})}{\partial \boldsymbol{\theta}_r}\right|_{\boldsymbol{\theta} = \widetilde{\boldsymbol{\theta}}}, \tag{4.37}$$

where $[\mathbf{I}_{\mathrm{F}}^{-1}(\widetilde{\boldsymbol{\theta}})]_{1:r,1:r}$ is the top-left $r \times r$ portion of the inverse of $\mathbf{I}_{\mathrm{F}}(\widetilde{\boldsymbol{\theta}})$, which is the value evaluated at $\boldsymbol{\theta} = \widetilde{\boldsymbol{\theta}}$ of the Fisher Information matrix

$$\mathbf{I}_{\mathrm{F}}(\boldsymbol{\theta}) = \begin{bmatrix} \mathbf{I}_{\mathrm{F}}(\boldsymbol{\theta})_{\boldsymbol{\theta}_r, \boldsymbol{\theta}_r} & \mathbf{I}_{\mathrm{F}}(\boldsymbol{\theta})_{\boldsymbol{\theta}_r, \boldsymbol{\theta}_s} \\ \mathbf{I}_{\mathrm{F}}(\boldsymbol{\theta})_{\boldsymbol{\theta}_s, \boldsymbol{\theta}_r} & \mathbf{I}_{\mathrm{F}}(\boldsymbol{\theta})_{\boldsymbol{\theta}_s, \boldsymbol{\theta}_s} \end{bmatrix} =$$

$$\begin{bmatrix} \mathbb{E}[\frac{\partial \ln f(\mathbf{x};\boldsymbol{\theta})}{\partial \boldsymbol{\theta}_r}(\frac{\partial \ln f(\mathbf{x};\boldsymbol{\theta})}{\partial \boldsymbol{\theta}_r})^T] & \mathbb{E}[\frac{\partial \ln f(\mathbf{x};\boldsymbol{\theta})}{\partial \boldsymbol{\theta}_r}(\frac{\partial \ln f(\mathbf{x};\boldsymbol{\theta})}{\partial \boldsymbol{\theta}_s})^T] \\ \mathbb{E}[\frac{\partial \ln f(\mathbf{x};\boldsymbol{\theta})}{\partial \boldsymbol{\theta}_s}(\frac{\partial \ln f(\mathbf{x};\boldsymbol{\theta})}{\partial \boldsymbol{\theta}_r})^T] & \mathbb{E}[\frac{\partial \ln f(\mathbf{x};\boldsymbol{\theta})}{\partial \boldsymbol{\theta}_s}(\frac{\partial \ln f(\mathbf{x};\boldsymbol{\theta})}{\partial \boldsymbol{\theta}_s})^T] \end{bmatrix}.$$

Here, we use $\boldsymbol{\theta} = [\boldsymbol{\theta}_r^T \ \boldsymbol{\theta}_s^T]^T$ to denote the concatenation of the primary parameter $\boldsymbol{\theta}_r$ (which is used to decide the hypothesis $H_0 : \boldsymbol{\theta}_r = \boldsymbol{\theta}_{r0}$ vs. $H_1 : \boldsymbol{\theta}_r \neq \boldsymbol{\theta}_{r0}$) and the nuisance parameter $\boldsymbol{\theta}_s$ which is needed to parameterize the likelihood function given by $f(\mathbf{x}; \boldsymbol{\theta})$. The value $\widehat{\boldsymbol{\theta}}_{s0}$ used in $\widetilde{\boldsymbol{\theta}} = [\boldsymbol{\theta}_{r0}^T \ \widehat{\boldsymbol{\theta}}_{s0}^T]^T$ is the ML estimate of $\boldsymbol{\theta}_s$ under H_0.

In our case, the primary and nuisance parameters are given as $\boldsymbol{\theta}_r = \mathrm{vec}(\overline{\boldsymbol{\Theta}})$ and $\boldsymbol{\theta}_s = [\gamma \quad \mathrm{vec}(\overline{\mathbf{R}})^T]^T$, so that we aim to decide between $H_0 : \boldsymbol{\theta}_r = \mathbf{0}$ vs. $H_1 : \boldsymbol{\theta}_r \neq \mathbf{0}$. From (4.9), we can see that the parts of $\ln f_{H_1}(\mathbf{X}, \mathbf{X}_s; \mathbf{R}, \gamma, \boldsymbol{\Theta})$ relevant to $\overline{\boldsymbol{\Theta}}$ can be written as

$$\frac{1}{\gamma}[\mathrm{vec}(\overline{\mathbf{V}}^T\overline{\mathbf{R}}^{-1}\overline{\mathbf{X}})^T\mathrm{vec}(\overline{\boldsymbol{\Theta}}) + \mathrm{vec}(\overline{\boldsymbol{\Theta}})^T\mathrm{vec}(\overline{\mathbf{V}}^T\overline{\mathbf{R}}^{-1}\overline{\mathbf{X}})-$$
$$\mathrm{vec}(\overline{\boldsymbol{\Theta}})^T(\mathbf{I} \otimes (\overline{\mathbf{V}}^T\overline{\mathbf{R}}^{-1}\overline{\mathbf{V}}))\mathrm{vec}(\overline{\boldsymbol{\Theta}})] \quad (4.38)$$

where we used the properties $\mathrm{tr}(\mathbf{A}^T\mathbf{B}) = \mathrm{vec}(\mathbf{A})^T\mathrm{vec}(\mathbf{B})$ and $\mathrm{tr}(\mathbf{A}^T\mathbf{B}\mathbf{A}) = \mathrm{vec}(\mathbf{A})^T(\mathbf{I}\otimes\mathbf{B})\mathrm{vec}(\mathbf{A})$, for two matrices $\mathbf{A} \in \mathbb{R}^{M\times N}$ and $\mathbf{B} \in \mathbb{R}^{M\times M}$. Then, ignoring any irrelevant scaling constants, we can say that for $f(\mathbf{x}; \boldsymbol{\theta}) = f_{H_1}(\mathbf{X}, \mathbf{X}_s; \mathbf{R}, \gamma, \boldsymbol{\Theta})$, we have

$$\frac{\partial \ln f(\mathbf{x}; \boldsymbol{\theta})}{\partial \boldsymbol{\theta}_r} = \frac{1}{\gamma}[\mathrm{vec}(\overline{\mathbf{V}}^T\overline{\mathbf{R}}^{-1}\overline{\mathbf{X}}) - (\mathbf{I} \otimes (\overline{\mathbf{V}}^T\overline{\mathbf{R}}^{-1}\overline{\mathbf{V}}))\mathrm{vec}(\overline{\boldsymbol{\Theta}})] \quad (4.39)$$

Since $\mathrm{vec}(\overline{\boldsymbol{\Theta}}) = \mathbf{0}$ under H_0, we can write

$$\mathbf{I}_{\mathbb{F}}(\boldsymbol{\theta})_{\boldsymbol{\theta}_r,\boldsymbol{\theta}_r} = \mathbb{E}[\frac{\partial \ln f(\mathbf{x}; \boldsymbol{\theta})}{\partial \boldsymbol{\theta}_r}(\frac{\partial \ln f(\mathbf{x}; \boldsymbol{\theta})}{\partial \boldsymbol{\theta}_r})^T] = \frac{1}{\gamma}[\mathbf{I} \otimes (\overline{\mathbf{V}}^T\overline{\mathbf{R}}^{-1}\overline{\mathbf{V}})], \quad (4.40)$$

where we used the fact that $\mathrm{vec}(\overline{\mathbf{V}}^T\overline{\mathbf{R}}^{-1}\overline{\mathbf{X}}) \sim \mathcal{N}(0, \gamma\mathbf{I} \otimes (\overline{\mathbf{V}}^T\overline{\mathbf{R}}^{-1}\overline{\mathbf{V}}))$ for H_0. Evaluating $\mathbf{I}_{\mathbb{F}}(\boldsymbol{\theta})_{\boldsymbol{\theta}_r,\boldsymbol{\theta}_r}$ at $\widetilde{\boldsymbol{\theta}} = [\boldsymbol{\theta}_{r0}^T \ \widehat{\boldsymbol{\theta}}_{s0}^T]^T$ using the ML estimate of $\boldsymbol{\theta}_s$ under H_0, we get

$$\mathbf{I}_{\mathbb{F}}(\widetilde{\boldsymbol{\theta}})_{\boldsymbol{\theta}_r,\boldsymbol{\theta}_r} = \frac{1}{\widetilde{\gamma}_0}[\mathbf{I} \otimes (\overline{\mathbf{V}}^T\widehat{\overline{\mathbf{R}}}_0^{-1}\overline{\mathbf{V}})], \quad (4.41)$$

where we let

$$\widehat{\overline{\mathbf{R}}}_0 = \begin{bmatrix} \widehat{\mathbf{R}}_0 & 0 \\ 0 & \widehat{\mathbf{R}}_0 \end{bmatrix}, \quad (4.42)$$

and

$$\widehat{\mathbf{R}}_0 = \frac{1}{\widehat{\gamma}_0}(\mathbf{X}_{Re}\mathbf{X}_{Re}^T + \mathbf{X}_{Im}\mathbf{X}_{Im}^T) + \mathbf{X}_{s,Re}\mathbf{X}_{s,Re}^T + \mathbf{X}_{s,Im}\mathbf{X}_{s,Im}^T, \qquad (4.43)$$

using the same estimated $\frac{1}{\widehat{\gamma}_0}$ as was used in the one-step GLRT statistic in (4.23). It can be shown that $\mathbf{I}_{\mathbb{F}}(\boldsymbol{\theta})_{\boldsymbol{\theta}_r,\boldsymbol{\theta}_s} = \mathbb{E}[\frac{\partial \ln f(\mathbf{x};\boldsymbol{\theta})}{\partial \boldsymbol{\theta}_r}(\frac{\partial \ln f(\mathbf{x};\boldsymbol{\theta})}{\partial \boldsymbol{\theta}_s})^T]$ is a null matrix, and hence

$$\left[\mathbf{I}_{\mathbb{F}}^{-1}(\widetilde{\boldsymbol{\theta}})\right]_{1:r,1:r} = \mathbf{I}_{\mathbb{F}}(\widetilde{\boldsymbol{\theta}})_{\boldsymbol{\theta}_r,\boldsymbol{\theta}_r}^{-1} = \widehat{\gamma}_0[\mathbf{I} \otimes (\overline{\mathbf{V}}^T\widehat{\overline{\mathbf{R}}}_0^{-1}\overline{\mathbf{V}})^{-1}]. \qquad (4.44)$$

Finally, evaluating $\frac{\partial \ln f(\mathbf{x};\boldsymbol{\theta})}{\partial \boldsymbol{\theta}_r}$ at $\boldsymbol{\theta} = \widetilde{\boldsymbol{\theta}}$, we can obtain the Rao test for the partially homogeneous environment from (4.37) as

$$\frac{1}{\widehat{\gamma}_0}\operatorname{tr}\{\overline{\mathbf{X}}^T\widehat{\overline{\mathbf{R}}}_0^{-1}\overline{\mathbf{V}}(\overline{\mathbf{V}}^T\widehat{\overline{\mathbf{R}}}_0^{-1}\overline{\mathbf{V}})^{-1}\overline{\mathbf{V}}^T\widehat{\overline{\mathbf{R}}}_0^{-1}\overline{\mathbf{X}}\}. \qquad (4.45)$$

4.3.7 Rao Test for the Fully Homogeneous Environment

For the fully homogeneous environment, the relevant quantities in (4.37) are the same as those in (4.39) and (4.41) but with γ and $\widehat{\gamma}_0$ set to one. The Rao test for the homogeneous environment is then given as

$$\operatorname{tr}\{\overline{\mathbf{X}}^T\widehat{\overline{\mathbf{R}}}_0^{-1}\overline{\mathbf{V}}(\overline{\mathbf{V}}^T\widehat{\overline{\mathbf{R}}}_0^{-1}\overline{\mathbf{V}})^{-1}\overline{\mathbf{V}}^T\widehat{\overline{\mathbf{R}}}_0^{-1}\overline{\mathbf{X}}\}, \qquad (4.46)$$

where $\widehat{\overline{\mathbf{R}}}_0$ is as given in (4.42) but with

$$\widehat{\mathbf{R}}_0 = \mathbf{X}_{Re}\mathbf{X}_{Re}^T + \mathbf{X}_{Im}\mathbf{X}_{Im}^T + \mathbf{X}_{s,Re}\mathbf{X}_{s,Re}^T + \mathbf{X}_{s,Im}\mathbf{X}_{s,Im}^T. \qquad (4.47)$$

4.3.8 Approximate Wald Test for the Partially Homogeneous Environment

The Wald test is given by the expression [55]

$$(\widehat{\boldsymbol{\theta}}_{r1} - \boldsymbol{\theta}_{r0})^T\left(\left[\mathbf{I}_{\mathbb{F}}^{-1}(\widehat{\boldsymbol{\theta}}_1)\right]_{1:r,1:r}\right)^{-1}(\widehat{\boldsymbol{\theta}}_{r1} - \boldsymbol{\theta}_{r0}), \qquad (4.48)$$

where $\widehat{\boldsymbol{\theta}}_1 = [\widehat{\boldsymbol{\theta}}_{r1}^T \ \widehat{\boldsymbol{\theta}}_{s1}^T]^T$ contains the concatenated ML estimates of $\boldsymbol{\theta}_r$ and $\boldsymbol{\theta}_s$ under H_1. Using (4.39), we can obtain the same value for $\mathbf{I}_{\mathbb{F}}(\boldsymbol{\theta})_{\boldsymbol{\theta}_r,\boldsymbol{\theta}_r}$ as in (4.40), since under H_1 we have $\mathrm{vec}(\overline{\mathbf{V}}^T\overline{\mathbf{R}}^{-1}\overline{\mathbf{X}}) \sim \mathcal{N}((\mathbf{I} \otimes (\overline{\mathbf{V}}^T\overline{\mathbf{R}}^{-1}\overline{\mathbf{V}}))\mathrm{vec}(\overline{\boldsymbol{\Theta}}), \gamma\mathbf{I} \otimes (\overline{\mathbf{V}}^T\overline{\mathbf{R}}^{-1}\overline{\mathbf{V}}))$. Moreover, from (4.39), we can say that the ML estimate of $\boldsymbol{\theta}_{r1} = \mathrm{vec}(\overline{\boldsymbol{\Theta}})$ can be given as

$$\widehat{\boldsymbol{\theta}}_{r1} = (\mathbf{I} \otimes \overline{\mathbf{V}}^T\overline{\mathbf{R}}^{-1}\overline{\mathbf{V}})^{-1}\mathrm{vec}(\overline{\mathbf{V}}^T\overline{\mathbf{R}}^{-1}\overline{\mathbf{X}}). \tag{4.49}$$

As in the case for the Rao test, it can be shown that $\mathbf{I}_{\mathbb{F}}(\boldsymbol{\theta})_{\boldsymbol{\theta}_r,\boldsymbol{\theta}_s}$ is a null matrix, so that $\left(\left[\mathbf{I}_{\mathbb{F}}^{-1}(\widehat{\boldsymbol{\theta}}_1) \right]_{1:r,1:r} \right)^{-1} = \mathbf{I}_{\mathbb{F}}(\boldsymbol{\theta})_{\boldsymbol{\theta}_r,\boldsymbol{\theta}_r}$. Then, evaluating (4.48) at the appropriate estimates $\widehat{\boldsymbol{\theta}}_{r1}$ and $\widehat{\boldsymbol{\theta}}_{s1}$, which we obtain using the heuristic in (4.12), we can arrive at an approximate Wald test for the partially homogeneous scenario as

$$\frac{1}{\widehat{\gamma}_1} \mathrm{tr}\{\overline{\mathbf{X}}^T\widehat{\overline{\mathbf{R}}}_1^{-1}\overline{\mathbf{V}}(\overline{\mathbf{V}}^T\widehat{\overline{\mathbf{R}}}_1^{-1}\overline{\mathbf{V}})^{-1}\overline{\mathbf{V}}^T\widehat{\overline{\mathbf{R}}}_1^{-1}\overline{\mathbf{X}}\}, \tag{4.50}$$

where

$$\widehat{\overline{\mathbf{R}}}_1 = \begin{bmatrix} \widehat{\mathbf{R}}_1 & 0 \\ 0 & \widehat{\mathbf{R}}_1 \end{bmatrix}, \tag{4.51}$$

and

$$\widehat{\mathbf{R}}_1 = \frac{1}{\widehat{\gamma}_1}(\widehat{\mathbf{Y}}_{Re}\widehat{\mathbf{Y}}_{Re}^T + \widehat{\mathbf{Y}}_{Im}\widehat{\mathbf{Y}}_{Im}^T) + \mathbf{X}_{s,Re}\mathbf{X}_{s,Re}^T + \mathbf{X}_{s,Im}\mathbf{X}_{s,Im}^T. \tag{4.52}$$

The $\widehat{\mathbf{Y}}_{Re}$ and $\widehat{\mathbf{Y}}_{Im}$ values used here are the same as those in (4.15) and (4.16), while $\widehat{\gamma}_1$ is the same as that used in the approximate GLRT.

4.3.9 Approximate Wald Test for the Fully Homogeneous Environment

For the fully homogeneous environment, the relevant quantities in (4.41), and (4.49) remain the same except with γ_0 set to one. An approximate Wald test for the homogeneous environment is then given as

$$\mathrm{tr}\{\overline{\mathbf{X}}^T\widehat{\overline{\mathbf{R}}}_1^{-1}\overline{\mathbf{V}}(\overline{\mathbf{V}}^T\widehat{\overline{\mathbf{R}}}_1^{-1}\overline{\mathbf{V}})^{-1}\overline{\mathbf{V}}^T\widehat{\overline{\mathbf{R}}}_1^{-1}\overline{\mathbf{X}}\}, \tag{4.53}$$

where $\widehat{\overline{\mathbf{R}}}$ is as given in (4.51), but with

$$\widehat{\mathbf{R}}_1 = \widehat{\mathbf{Y}}_{Re}\widehat{\mathbf{Y}}_{Re}^T + \widehat{\mathbf{Y}}_{Im}\widehat{\mathbf{Y}}_{Im}^T + \mathbf{X}_{s,Re}\mathbf{X}_{s,Re}^T + \mathbf{X}_{s,Im}\mathbf{X}_{s,Im}^T. \qquad (4.54)$$

4.4 Numerical Simulations

The performance of the proposed detectors utilizing spectral symmetry was compared to those of their counterparts which did not utilize this constraint. The experimental signal model followed (4.1), and for the partially homogeneous case γ chosen in each MC simulation as a uniformly distributed random variable between one and three. Following similar modeling choices in [65, 63, 7], and [67], the covariance matrix $\mathbf{R}$ was set so that its i,j th element was given as $\rho^{|i-j|}$, with $\rho = 0.95$. We formed $\mathbf{V}$ from orthonormalized length-N steering vectors of the form $\mathbf{v} = [1 \ \exp\{-2\pi i\omega\} \ \dots \ \exp\{-2\pi i(N-1)\omega\}]^H$, where i is the imaginary unit, at angles equally spaced from $\omega = 0$ to $\omega = 0.3$ in r increments. Each element of the variable $\mathbf{\Theta} \in \mathbb{C}^{r \times K_p}$ was chosen as $CN(0,1)$, and $\mathbf{\Theta}$ was multiplied by a constant scaling set to achieve a desired SINR, defined as $\frac{1}{K_p}\text{tr}\left(\mathbf{\Theta}^H\mathbf{V}^H\mathbf{R}^{-1}\mathbf{V}\mathbf{\Theta}\right)$. The false alarm rate was set as 10^{-3}, and the thresholds to be used were set using 4×10^4 Monte-Carlo (MC) simulations, while 2×10^3 MC simulations were used for calculating the probability of detection across SINR. In all of the shown plots, it is evident that the incorporation of spectral symmetry gives a significant improvement in detection performance. In particular, the approximate one-step GLR, though derived in a heuristic manner, gives an improved detection performance as compared to its two-step counterpart, and can still be implemented in a closed-form, non-iterative manner. Nonetheless, the overall best performing detector across the various settings, as suggested by the presented plots, is the spectrally symmetric Rao test, which may be the best choice of detector due to its high-level performance and its low computational cost in not needing to compute the covariance estimate for the H_1 case.

4.5 Conclusion

Adaptive detectors were derived under the assumption of spectrally symmetric noise under the framework of the GLR, Rao, and Wald test procedures. In particular, newly reported detectors were given for approximate one-step GLR and Wald tests for the homogeneous and partially homogeneous environments, and detectors were given for the range-spread and subspace signal cases, which are widely utilized in adaptive radar. It was shown through numerical simulations that the incorporation of the spectral symmetry constraint, which is motivated by real-world measurements of ground clutter reflections, allows the design of detectors with vastly improved performance over their counterparts which do not incorporate this constraint.

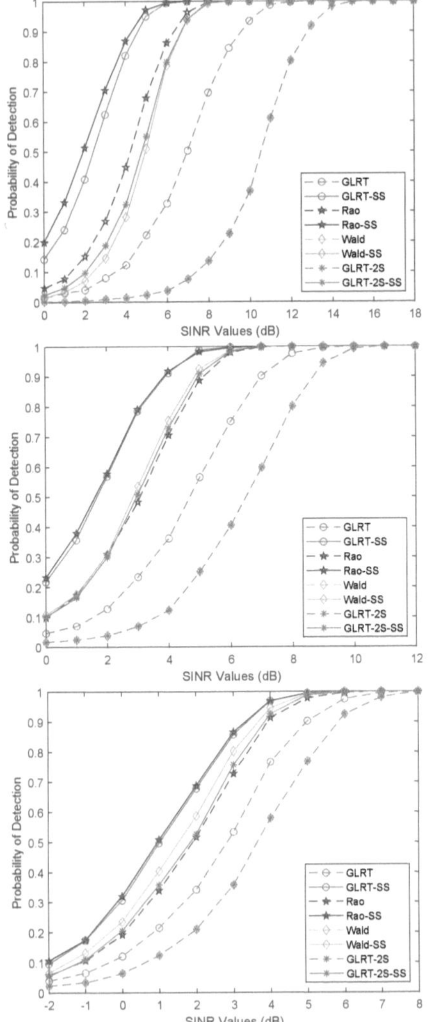

Figure 4.1: Probabilities of detection versus SINR of the various detectors in the homogeneous scenario, for $K_p = 1$, $r = 4$, $N = 16$, and, from top to bottom, $K_s = 24$, 32, and 48.

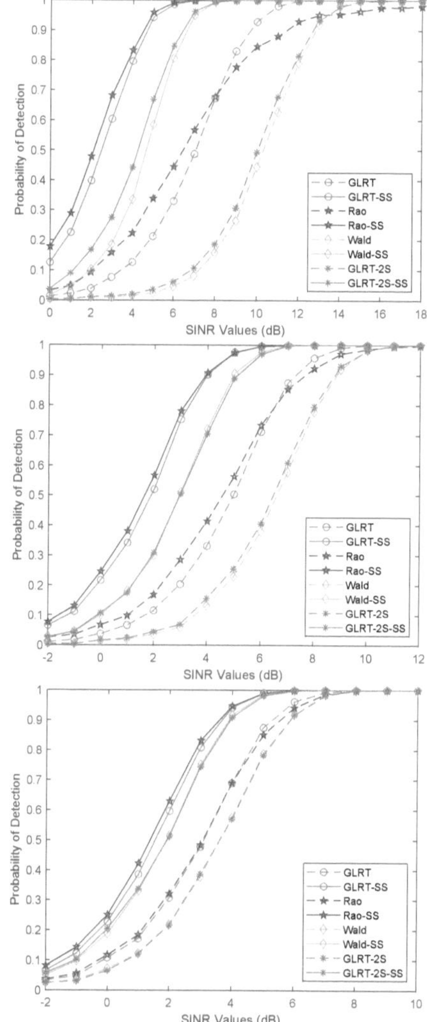

Figure 4.2: Probabilities of detection versus SINR of the various detectors in the partially homogeneous scenario, for $K_p = 1$, $r = 4$, $N = 16$, and, from top to bottom, $K_s = 24$, 32, and 48.

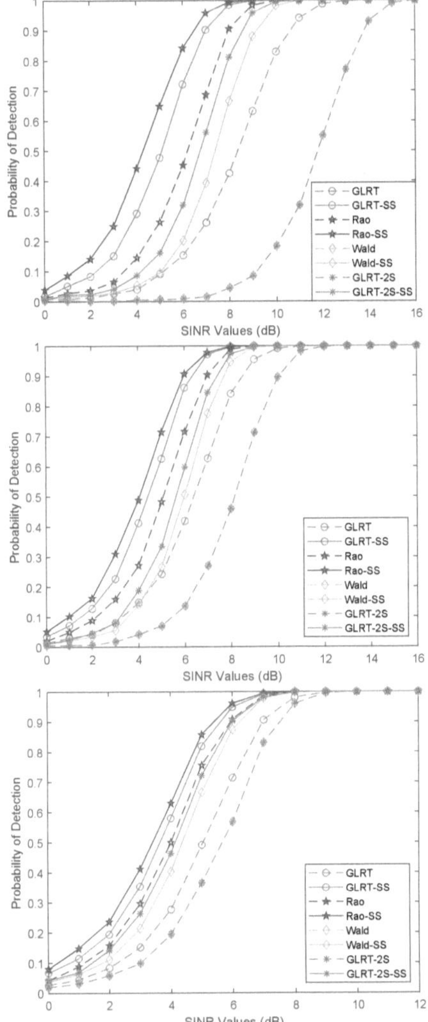

Figure 4.3: Probabilities of detection versus SINR of the various detectors in the homogeneous scenario, for $K_p = 16$, $r = 4$, $N = 16$, and, from top to bottom, $K_s = 24$, 32, and 48.

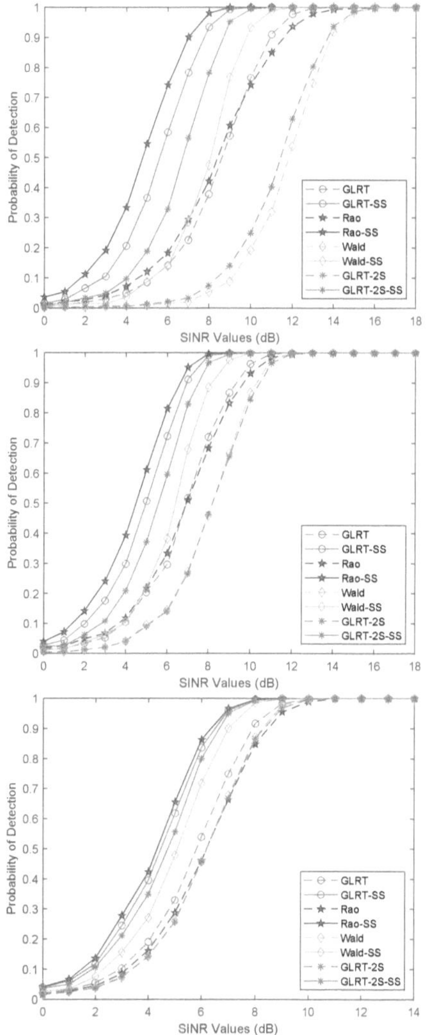

Figure 4.4: Probabilities of detection versus SINR of the various detectors in the partially homogeneous scenario, for $K_p = 16$, $r = 4$, $N = 16$, and, from top to bottom, $K_s = 24$, 32, and 48.

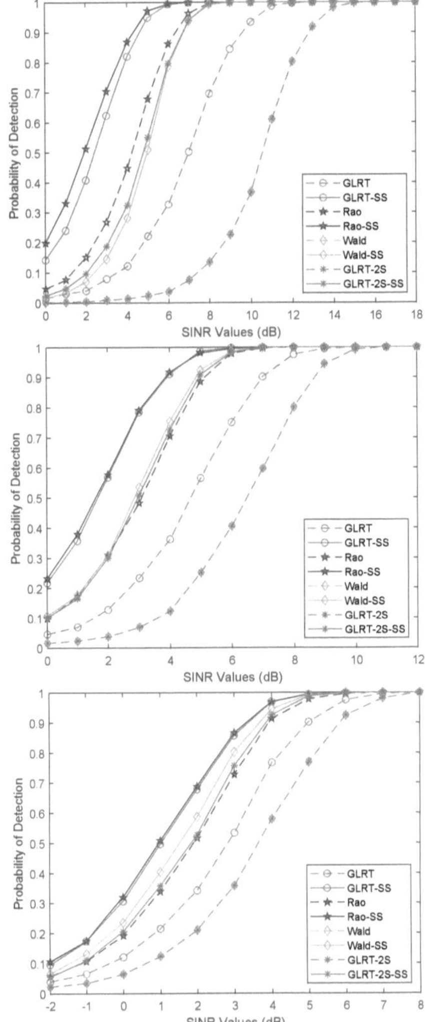

Figure 4.5: Probabilities of detection versus SINR of the various detectors in the homogeneous scenario, for $K_p = 16$, $r = 1$, $N = 16$, and, from top to bottom, $K_s = 24$, 32, and 48.

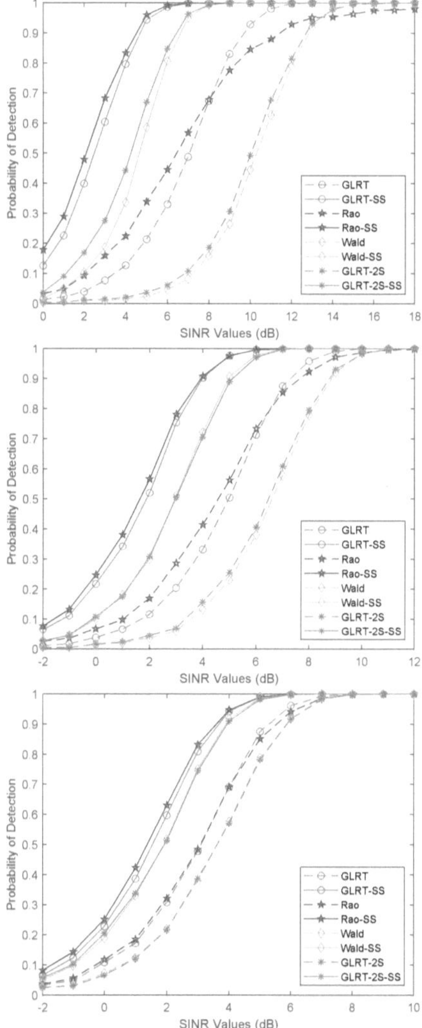

Figure 4.6: Probabilities of detection versus SINR of the various detectors in the partially homogeneous scenario, for $K_p = 16$, $r = 1$, $N = 16$, and, from top to bottom, $K_s = 24$, 32, and 48.

Chapter 5: Adaptive Detection in the Presence of Spectral Symmetry and the Absence of Secondary Data

5.1 Overview

In this chapter, different from the others, there is no secondary data used to estimate the covariance matrix. Only primary data is available, which may or may not contain the signal of interest. Moreover, the signal itself is also modeled as being random, as opposed to being modeled as a deterministic unknown as in the other chapters. The detectors introduced in this section for the SIRV case are given in an ad-hoc manner, but are based off of relevant results in the literature. It is shown that spectral symmetry and canonical correlation analysis can be utilized to give robust adaptive detection even in the absence of secondary data.

5.2 Introduction

One of the ongoing challenges in adaptive detection is the scarcity of secondary data for estimating the covariance matrix of the colored Gaussian noise. In applications like adaptive radar [91, 47], where noise is often used to model both thermal noise present in the receiver as well as unwanted clutter returns, this scarcity is due to the fact that secondary data is often heterogeneous, and does not share the same statistical distribution as the noise present in the primary data. In this regard, it is desirable to derive detectors which achieve an acceptable performance using a limited amount of secondary data.

An important factor in the design of detectors for data-limited scenarios is incorporation of additional knowledge of the secondary data. One of the key characteristics of measured ground clutter data [16] which has received attention in the literature is that of spectral symmetry, meaning that the real and imaginary parts of such clutter returns can be approximated as being statistically independent. It can be shown that this property (which does not necessarily hold for arbitrary circular complex Gaussian data [38]) implies that the unknown covariance matrix of the colored noise has all real elements, which can greatly enhance its estimation. A number of papers [31, 49, 50, 34, 35] have considered the case of adaptive detection when the additive noise is known to be spectrally symmetric. However, all of these papers considered the case of point-like targets that are present in one range cell and belong to the subspace spanned by a single steering vector. In practical applications, it is often of interest to detect a signal which can lie along a subspace spanned by several vectors [57, 67, 68, 19, 20], which is a modeling choice often made to account for possible uncertainties in the steering vector, for targets which may span multiple angle or Doppler values, or for detection across a certain swath of the Doppler/azimuth grid. In addition, range-spread modeling of the target signal is often useful for newer, higher-resolution radar systems, in which a potential target could be spread out over multiple range-cells which then need to be tested for target presence simultaneously [24]. Moreover, [57, 67, 68, 19, 20], like most other papers in the adaptive detection literature, rely on the presence of secondary, target-free data to estimate the covariance matrix of the additive noise. While incorporation of constraints like spectral symmetry and the like can help reduce the requirements on the sample size of the secondary data, it is of interest to also consider the case of no secondary data being available, as was assumed, e.g., in [41]. Furthermore, it is also of interest to consider the case when the target subspace is unknown except for its dimension; such a signal model can account for a high degree of uncertainty in the target steering vectors, and was considered in various works like [77], [23], and [69].

Toward this end, in this paper we propose an ad-hoc detector based on canonical correlations analysis (CCA) [70, 73] to test for the presence or absence of a range-spread target, which is modeled as having an unknown signature with Gaussian (or non-Gaussian) coordinates in a high-dimensional space, and contaminated with Gaussian or non-Gaussian (such as with spherically invariant random vector (SIRV) [25, 26, 42, 22]) noise. No secondary data is assumed to be available. Such a detection approach can be viewed as an extension of the problem in [81], which considered passive MIMO radar detection with a reference signal and Gaussian noise, to the case of an active radar scenario in which the additive noise (but not necessarily the signal coordinates) are spectrally symmetric random variables. Another possible application for our proposed detection approach is the case of passive MIMO or multichannel radar with no reference signal but with spectrally symmetric noise.

The main principle in the design of our proposed detector is that the cross-correlation between the real and imaginary components of the received data should only be nonzero if a target reflection is present. By incorporating this principle in a framework based on canonical correlation analysis, we will propose a detection scheme which will be analytically shown to have a constant false alarm rate (CFAR) with respect to the unknown covariance matrix for the case of Gaussian noise and empirically shown to be invariant with respect to both the covariance and textures (unknown scalings across range cells) in the case of SIRV noise.

5.3   Detection Under Spectral Symmetry

5.3.1   Problem Setup

The detection problem we are interested in can be described as a hypothesis test of the form

$$\begin{cases} H_1 : \mathbf{X} = \mathbf{V}\Theta + \mathbf{N}, \\ H_0 : \mathbf{X} = \mathbf{N}, \end{cases} \tag{5.1}$$

where $\mathbf{V} \in \mathbb{C}^{N \times r}$, $1 \leq r < N$, is an unknown matrix whose columns span the signal subspace, and $\Theta \in \mathbb{C}^{r \times K_p}$ represents the unknown coordinates of the hypothetical signal within this subspace, with the number of received data samples being $K_p \geq r$. We assume that $\Theta \stackrel{c}{\sim} \mathcal{CN}(\mathbf{0}, \mathbf{I})$, where we use $\stackrel{c}{\sim}$ to mean "columnwise distributed as"; note that if the covariance matrix of Θ was some arbitrary matrix $\mathbf{M}$, then we could simply redefine $\mathbf{V}$ and Θ as $\mathbf{V}_{new} = \mathbf{M}^{\frac{1}{2}}\mathbf{V}$ and $\Theta_{new} = \mathbf{M}^{-\frac{1}{2}}\Theta$, where clearly $\Theta_{new} \stackrel{c}{\sim} \mathcal{CN}(\mathbf{0}, \mathbf{I})$. The matrix $\mathbf{X}$ represents the CUT, and no secondary signal-free training data are assumed to be present. Finally, we assume that the additive noise $\mathbf{N}$ is distributed as $\mathbf{N} \stackrel{c}{\sim} \mathcal{CN}(\mathbf{0}, \mathbf{R}_n)$, with $\mathbf{R}_n \in \mathbb{R}^{N \times N}$, and is statistically independent of the signal Θ.

For the present case, we model the real and imaginary portions of the received signal at the surveillance channel as

$$\mathbf{X}_{Re} = \beta(\mathbf{V}_{Re}\Theta_{Re} - \mathbf{V}_{Im}\Theta_{Im}) + \mathbf{N}_{Re} \tag{5.2}$$

and

$$\mathbf{X}_{Im} = \beta(\mathbf{V}_{Re}\mathbf{\Theta}_{Im} + \mathbf{V}_{Im}\mathbf{\Theta}_{Re}) + \mathbf{N}_{Im}, \tag{5.3}$$

respectively, where we use the subscripts "Re" and "Im" to denote the real and imaginary portions of the relevant variables, and $\beta = 0$ for H_0 while $\beta \neq 0$ for H_1. It can readily be inferred that in the case of H_1, we have $\mathbb{E}[\mathbf{X}_{Re}\mathbf{X}_{Re}^T] = \mathbb{E}[\mathbf{X}_{Im}\mathbf{X}_{Im}^T] = \mathbf{R}_1 = \beta^2\mathbf{V}_{Re}\mathbf{V}_{Re}^T + \beta^2\mathbf{V}_{Im}\mathbf{V}_{Im}^T + \mathbf{R}_n$, so that the covariance matrix of the concatenated received signal $\overline{\mathbf{X}} = [\mathbf{X}_{Re}^T\ \mathbf{X}_{Im}^T]^T$ can be written as

$$\overline{\mathbf{R}}_1 = \begin{bmatrix} \mathbf{R}_1 & \beta^2\mathbf{V}_{Re}\mathbf{V}_{Im}^T \\ \beta^2\mathbf{V}_{Im}\mathbf{V}_{Re}^T & \mathbf{R}_1 \end{bmatrix}. \tag{5.4}$$

On the other hand, for H_0, we have $\mathbb{E}[\mathbf{X}_{Re}\mathbf{X}_{Re}^T] = \mathbb{E}[\mathbf{X}_{Im}\mathbf{X}_{Im}^T] = \mathbf{R}_n$, so that the covariance matrix of $\overline{\mathbf{X}}$ then becomes

$$\overline{\mathbf{R}}_0 = \begin{bmatrix} \mathbf{R}_n & \mathbf{0} \\ \mathbf{0} & \mathbf{R}_n \end{bmatrix}. \tag{5.5}$$

The fundamental principle behind CCA-based detection in this setting is to try to distinguish the structure of the covariance matrix of $\overline{\mathbf{X}}$ between $\overline{\mathbf{R}}_0$ for H_0 and $\overline{\mathbf{R}}_1$ for H_1, which is equivalent to distinguishing whether $\beta = 0$ or $\beta \neq 0$. This attempt at distinguishing between $\overline{\mathbf{R}}_0$ and $\overline{\mathbf{R}}_1$ will be done in a heuristic, ad-hoc manner in the next section.

5.3.2 Detection Based on CCA

The CCA-based detector we use for the spectrally symmetric scenario, a similar form of which was derived in [81] as the GLRT for a different scenario, is given as

$$t_{hom} = \prod_{i=1}^{r} \frac{1}{(1-k_i^2)} \underset{H_0}{\overset{H_1}{\gtrless}} \eta. \tag{5.6}$$

Here, η is a detection threshold chosen to satisfy a preset false alarm rate, and k_i are the singular values of the matrix $\mathbf{S}_{ss}^{-\frac{1}{2}}\mathbf{S}_{sr}\mathbf{S}_{rr}^{-\frac{1}{2}}$, where in the general scenario $\mathbf{S}_{ss} =$

$\frac{1}{K_p}\mathbf{X}_s\mathbf{X}_s^H$, $\mathbf{S}_{rr} = \frac{1}{K_p}\mathbf{X}_r\mathbf{X}_r^H$, and $\mathbf{S}_{sr} = \frac{1}{K_p}\mathbf{X}_s\mathbf{X}_r^H$ are the maximum likelihood estimates of $\frac{1}{K_p}\mathbb{E}[\mathbf{X}_s\mathbf{X}_s^H]$, $\frac{1}{K_p}\mathbb{E}[\mathbf{X}_r\mathbf{X}_r^H]$, and $\frac{1}{K_p}\mathbb{E}[\mathbf{X}_s\mathbf{X}_r^H]$, respectively. [4] In the general scenario, it is of interest to detect the presence or absence of correlation between some received data matrices $\mathbf{X}_s$ and $\mathbf{X}_r$; in our case, the correlation of interest to be detected is between the real and imaginary components $\mathbf{X}_{Re}$ and $\mathbf{X}_{Im}$ of the received data $\mathbf{X}$.

Due to this fact, the k_i terms used in (5.6) for the spectrally symmetric case can be considered as the singular values of the matrix $\mathbf{S}^{-\frac{1}{2}}\mathbf{S}_{cc}\mathbf{S}^{-\frac{1}{2}}$, where $\mathbf{S} = \frac{1}{2K_p}(\mathbf{X}_{Re}\mathbf{X}_{Re}^T + \mathbf{X}_{Im}\mathbf{X}_{Im}^T)$ and $\mathbf{S}_{cc} = \frac{1}{K_p}\mathbf{X}_{Re}\mathbf{X}_{Im}^T$. Noting that $\mathbf{S} \sim \mathcal{W}(2K_p, \mathbf{R})$, we can show that the statistic in (5.6) has a constant false alarm rate (CFAR) with respect to $\mathbf{R}$. This can be done by noting that the denominator in (5.6) consists of the first r terms of the overall product $\prod_{i=1}^{N}(1 - k_i^2)$, which can be written equivalently as

$$
\begin{aligned}
\prod_{i=1}^{N}(1 - k_i^2) &= \det(\mathbf{I} - \mathbf{S}^{-\frac{1}{2}}(\mathbf{X}_{Re}\mathbf{X}_{Im}^T)\mathbf{S}^{-1}(\mathbf{X}_{Im}\mathbf{X}_{Re}^T)\mathbf{S}^{-\frac{1}{2}}) \\
&= \frac{\det(\mathbf{S} - \mathbf{X}_{Re}\mathbf{X}_{Im}^T\mathbf{S}^{-1}\mathbf{X}_{Im}\mathbf{X}_{Re}^T)}{\det(\mathbf{S})} \\
&= \frac{\det(\mathbf{R}^{-1})\det(\mathbf{S} - \mathbf{X}_{Re}\mathbf{X}_{Im}^T\mathbf{S}^{-1}\mathbf{X}_{Im}\mathbf{X}_{Re}^T)}{\det(\mathbf{R}^{-1})\det(\mathbf{S})} \\
&= \frac{\det(\widetilde{\mathbf{S}} - \widetilde{\mathbf{X}}_{Re}\widetilde{\mathbf{X}}_{Im}^T\widetilde{\mathbf{S}}^{-1}\widetilde{\mathbf{X}}_{Im}\widetilde{\mathbf{X}}_{Re}^T)}{\det(\widetilde{\mathbf{S}})},
\end{aligned}
\tag{5.7}
$$

where $\widetilde{\mathbf{X}}_{Re} = \mathbf{R}^{-\frac{1}{2}}\mathbf{X}_{Re}$, $\widetilde{\mathbf{X}}_{Im} = \mathbf{R}^{-\frac{1}{2}}\mathbf{X}_{Im}$, and $\widetilde{\mathbf{S}} = \mathbf{R}^{-\frac{1}{2}}\mathbf{S}\mathbf{R}^{-\frac{1}{2}}$. Since $\widetilde{\mathbf{X}}_{Re}$ and $\widetilde{\mathbf{X}}_{Im}$ are both white Gaussian, and $\widetilde{\mathbf{S}}$ is white Wishart, we can say from (5.7) that the detector given in (5.6) consists only of terms that are independent of $\mathbf{R}$, and is hence CFAR with respect to it.

---

[4] To reduce the notational burden, we will use expressions like $\mathbf{X}_s$ and $\mathbf{X}_r$ to refer to both the random variable and its sample realization, where the difference should be clear from the context.

5.3.3   Extension to the Case of SIRV Noise

Up to this point, we have assumed that the noise present in our detection problem is Gaussian distributed. This is a widely used assumption throughout the adaptive detection literature (see, e.g., [56, 80, 59, 67, 68, 19, 20], and many others), and provides a simple approach with elegant mathematical tractability. However, in certain applications like adaptive radar, measured data from ground clutter returns have been found to be better approximated by certain non-Gaussian distributions, most commonly those described by the SIRV model (e.g.,[25, 26, 42, 22]). Thus, it is of interest to also consider the case of additive noise in which the kth column of $\mathbf{N}$ in (5.1) is now of the form $\sqrt{\tau_k}\mathbf{n}_k$, where $\mathbf{n}_k$ is distributed as before and τ_k is an unknown scaling value chosen independently for each column. In the design of our heuristic detector, we will treat the τ_k values as deterministic unknowns. In order to enable the application of a detector based on CCA for our scenario, we will extend the τ_k scaling so that each column of $\mathbf{X}$ is now given as $\sqrt{\tau_k}\mathbf{x}_i$. In this new setting, we will modify the estimate $\mathbf{S}$ using the fixed-point estimator [42] whose $(i+1)$th iteration is given as

$$\mathbf{S}(i+1) = \frac{N}{K}\sum_{k=1}^{K}\frac{1}{\mathbf{x}_k^T\mathbf{S}(i)^{-1}\mathbf{x}_k}\mathbf{x}_k\mathbf{x}_k^T, \tag{5.8}$$

where we initialize the estimate as $\mathbf{S}(1) = \mathbf{I}$, and where $\mathbf{x}_k$ is the kth column of the concatenated matrix $\mathbf{X}_t = [\mathbf{X}_{Re}\ \mathbf{X}_{Im}]$. Denoting the final estimate of $\mathbf{S}$ as $\check{\mathbf{S}}$, the estimate of the cross-correlation is now given as

$$\check{\mathbf{S}}_{cc} = \frac{N}{K}\sum_{k=1}^{\frac{K}{2}}\sqrt{\frac{1}{\mathbf{x}_k^T\check{\mathbf{S}}^{-1}\mathbf{x}_k}}\mathbf{x}_k\mathbf{x}_{k+K}^T\sqrt{\frac{1}{\mathbf{x}_{k+K}^T\check{\mathbf{S}}^{-1}\mathbf{x}_{k+K}}}. \tag{5.9}$$

An ad-hoc CCA-based detection statistic can be used for the SIRV case in the form of the detector in (5.6), except with the k_i variables now being the singular values of the matrix $\breve{\mathbf{S}}^{-\frac{1}{2}}\breve{\mathbf{S}}_{cc}\breve{\mathbf{S}}^{-\frac{1}{2}}$.

5.4   Numerical Simulations

Here, we present a brief overview of our proposed detection scheme for signals in spectrally symmetric noise that is both Gaussian (denoted as "CCA-hom" in the plots) and SIRV (denoted as "CCA-nonhom"). Both schemes use the detection structure given in (5.6). For comparison purposes, we also included two other detectors which would be reasonable to implement in the given scenario. The first of these is the cross-correlation detector, which was also used as a comparison in [81], given as

$$t_{cross-corr} = \operatorname{tr}\left(\mathbf{X}_{Re}\mathbf{X}_{Im}^{T}\mathbf{X}_{Im}\mathbf{X}_{Re}^{T}\right). \tag{5.10}$$

The cross-correlation detector is a widely used tool in passive radar signal processing, and can be easily implemented without any matrix inversion. However, it can be seen by inspection that this detector is not CFAR with respect to the covariance matrix $\mathbf{R}$ and, in the nonhomogeneous case, to the scalings τ_k. The other detector we compare against is the even more basic energy detector, given as

$$t_{energy-det} = \operatorname{tr}\left(\mathbf{X}^{T}\mathbf{X}\right). \tag{5.11}$$

The energy detector is also non-CFAR with respect to $\mathbf{R}$ and τ_k. In contrast to the cross-correlation detector, which measures the degree of correlation between $\mathbf{X}_{Re}$ and $\mathbf{X}_{Im}$, the energy detector only measures the energy present in the sample covariance matrix $\mathbf{X}\mathbf{X}^{T}$, which obviously should be greater under H_1 than H_0. Thus, it can be argued that the energy detector does not utilize the prior assumption of spectral symmetry to the same degree as the cross-correlation detector does. As will be seen in the simulations, the greater scenario knowledge incorporated by the cross-correlation detector gives it a small improvement over the energy detector in homogeneous environments. However, it will also be seen that the energy detector is more robust to the data heterogeneity that occurs when τ_k are non-unity, giving it a slight edge over the cross-correlation detector in heterogeneous environments.

Figure 5.1 shows the performance of the homogeneous and nonhomogeneous CCA detectors for the case when the underlying data is homogeneous, meaning that the scalings τ_k are all ones. Similarly to [81], we performed Monte-Carlo (MC) simulations where at each iteration we set the matrix $\mathbf{R}_n$ as $\mathbf{R}_n = \mathbf{A}^H \mathbf{A}$, where we set the entries of $\mathbf{A} \in \mathbb{C}^{8 \times 8}$ as $\mathcal{N}(0,1)$ random samples. The entries of $\mathbf{V} \in \mathbb{C}^{8 \times 1}$ were set as $\mathcal{CN}(0,1)$ random samples, and were scaled to provide SINR values at the reference and surveillance channels defined as $10\log_{10}(\frac{\text{tr}(\mathbf{V}^H \mathbf{V})}{\text{tr}(\mathbf{R}_n)})$. The number of data samples used was varied across $K = 16$, $K = 24$, and $K = 40$. The SINR was swept across a range of values to devise the P_d (probability of detection) versus SINR plots. The false alarm rate was kept at $P_{fa} = 10^{-3}$, where the threshold was chosen via 3×10^4 MC simulations, and 3×10^3 MC simulations were used to make the P_d vs SINR plots. For the SIRV-based detector, five iterations were used for the estimation procedure in (5.8). From the plots in Figure 5.1, we see that the cross-correlation detector and energy detector outperform the CCA-based detectors for low training data scenarios. As was mentioned earlier, and as will later be shown experimentally, this advantage in performance comes at the cost of not having the important CFAR property. With regards to the CCA-based detectors, we see that the homogeneous CCA detector outperforms the nonhomogeneous one, which is reasonable since that is the scenario it was designed for. Nonetheless, the performance of the nonhomogeneous CCA detector is still quite close to the homogeneous one.

Figure 5.2 shows the performance of the homogeneous and nonhomogeneous CCA detectors for the case when the underlying data is nonhomogeneous, again for an empirical false alarm rate of 10^{-3}. Specifically, the scalings τ_k were chosen using the MATLAB function "gamrnd" (used for generating gamma-distributed random samples) with parameters 1 and 0.5, inspired by the K-distributed clutter model used in seminal radar papers like [42] and observed in real-world measured data. For comparison purposes, to ease the effect of heterogeneity on the homogeneous CCA detector, we smoothed out the τ_k scalings using a moving-average filter of length five. However, as can be seen in Figure 5.2, even with this smoothing present the homogeneous CCA detector suffers a severe performance loss as compared to its more robust heterogeneous version. In simulations not shown here, we have observed that when the smoothing is not applied to the scalings, the performance of homogeneous CCA degrades even further and the detector fails completely. The energy detector is seen to give slightly better performance than the cross-correlation detector for $K = 16$ and $K = 24$, and is only marginally worse than the cross-correlation detector for $K = 40$, which suggests that it may be more robust in heterogeneous, low training data scenarios than the cross-correlation detector.

When we introduced the ad-hoc CCA-based detector for the heterogeneous environment, we made an assumption that the signal component was also scaled by the τ_k values in order to more readily incorporate results from SIRV covariance estimation theory. In order to examine the performance of our ad-hoc approach when the signal is actually homogeneous and Gaussian, and only the additive noise is of an SIRV nature, we formed the plots given in Figure 5.3. The results given in these plots suggest that in this scenario, the homogeneous CCA-based detector fails completely, while the nonhomogeneous CCA-based detector outperforms the energy detector and cross-correlation detector in all but the high-SINR, low training data cases.

Finally, in Figure 5.4, we plotted the thresholds (divided by the sum across all α values for each detector) needed for the homogeneous and nonhomogeneous CCA detectors to maintain a false alarm rate of 10^{-3}. In this case, the scalings τ_k corresponded to entries from the vector $(1-\alpha)\mathbf{1} + \alpha\boldsymbol{\gamma}$, where $\boldsymbol{\gamma}$ was a vector of random variables sampled using the aforementioned "gamrnd" with parameters 1 and 0.5, and α was a scalar swept from zero (fully homogeneous) to one (fully heterogeneous). Reasonably, from the plots in Figure 5.4 we see that for low values of α the thresholds for homogeneous and nonhomogeneous CCA detectors coincide, but the threshold for the homogeneous version begins to diverge significantly for larger α values, showing an empirical lack of CFAR of the homogeneous CCA detector with respect to the data scalings. Since the CFAR property is highly sought after in practical radar systems, the presented results suggest that a hetereogeneous CCA detector may be more desirable than its homogeneous counterpart in the presence of spiky real-world clutter returns, which may otherwise flood the system with false alarms.

In Figure 5.5, we plotted the normalized thresholds for a false alarm rate of 10^{-3} when the τ_k were drawn the same way as described for Figure 5.2. This time, the α values were swept in the same manner as in Figure 5.4, but now it was the underlying noise covariance matrix which was altered as $\mathbf{R}_n = (1-\alpha)\mathbf{R}_a + \alpha\mathbf{R}_b$, where $\mathbf{R}_a$ was set the same way as $\mathbf{R}_n$ in Figure 5.1, and $\mathbf{R}_b$ was a tapered matrix whose i,jth element was given as $\rho^{|i-j|}$, with $\rho = 0.95$. Reasonably, in the heterogeneous case only the

SIRV-based CCA detector had a constant threshold across the α values, while the other detectors swayed around to various degrees, needing to adjust their thresholds to maintain the required false alarm rate.

5.5 Conclusion

We considered a detection scenario in which we were interested in ascertaining the presence or absence of a signal with an unknown signature in the presence of Gaussian or SIRV noise with an unknown covariance matrix. By incorporating the assumption of spectral symmetry, we were able to transform the problem to one of determining the degree of correlation between the real and imaginary components of the received data samples. To the best of the author's knowledge, this is the first time that CCA was used along with spectral symmetry in order to alleviate the requirement for secondary data. It was shown that a detection approach based on CCA gave a reasonable performance in this environment, and with some modification to the covariance and cross-covariance estimates, also gave adequate performance in the case of SIRV noise. Both detectors were shown to be robust with respect to changes in the covariance matrix, with the modified SIRV version also being robust to arbitrary scalings of the received data.

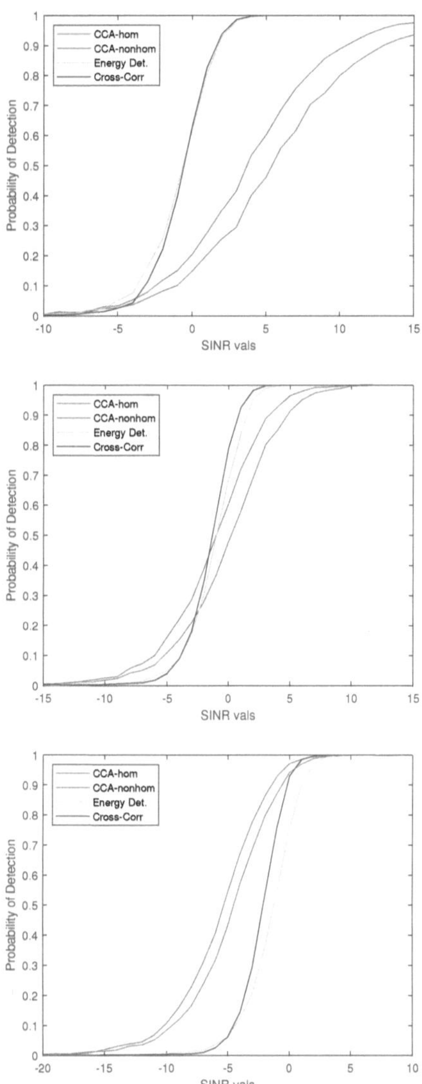

Figure 5.1: Performance of CCA and other detectors for the homogeneous scenario, with $N = 8$, $r = 3$, and $K = 16$ (top), $K = 24$ (middle), and $K = 40$ (bottom).

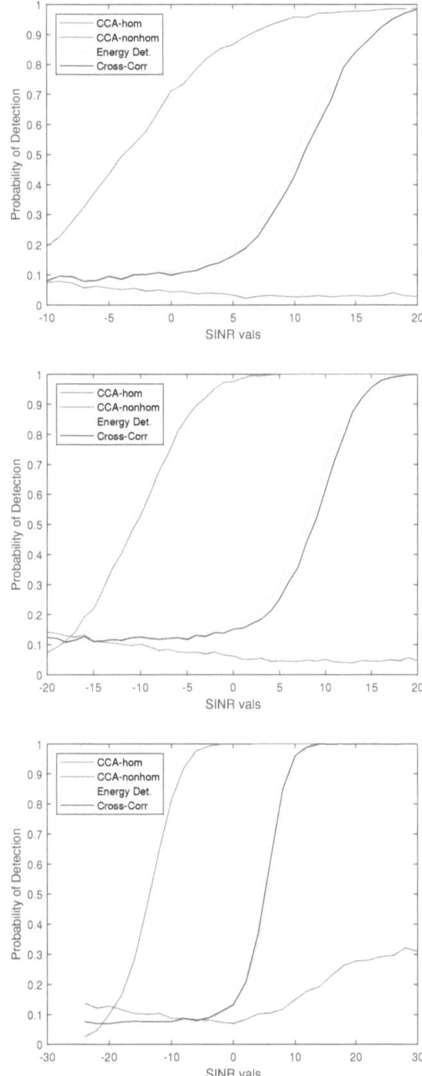

Figure 5.2: Performance of CCA and other detectors for the nonhomogeneous scenario, with $N = 8$, $r = 3$, and $K = 16$ (top), $K = 24$ (middle), and $K = 40$ (bottom).

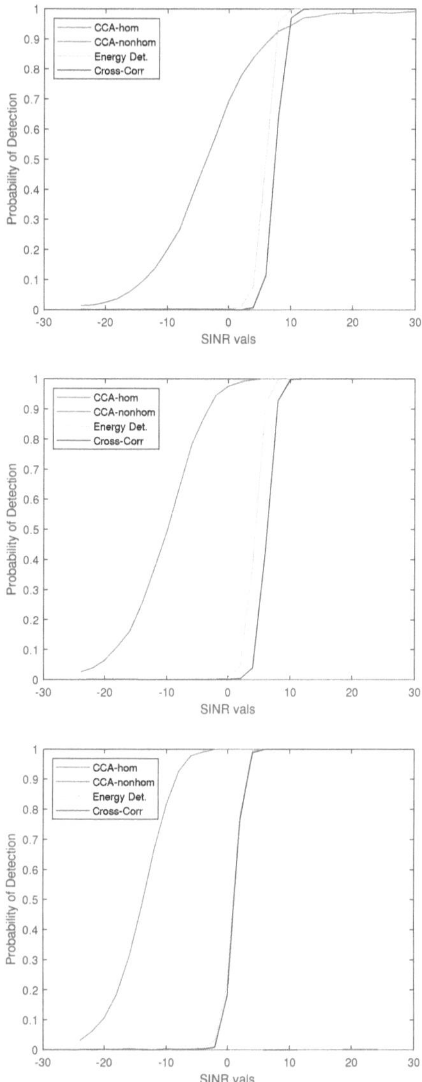

Figure 5.3: Performance of CCA and other detectors for the nonhomogeneous scenario, when only the additive noise is scaled, with $N = 8$, $r = 3$, and $K = 16$ (top), $K = 24$ (middle), and $K = 40$ (bottom).

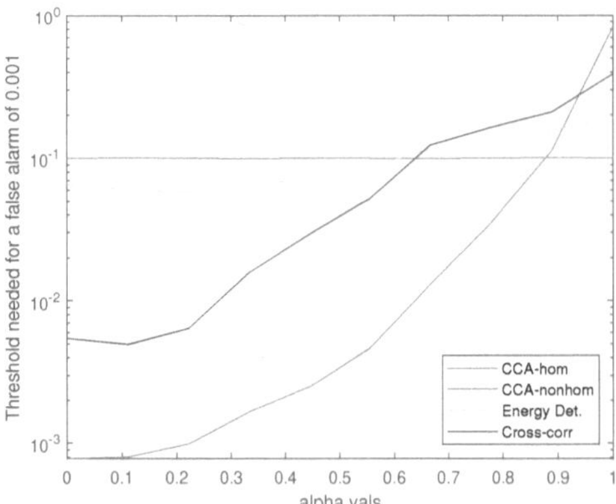

Figure 5.4: Required threshold to maintain $P_{fa} = 10^{-3}$ for the CCA-based and other detectors across various values of α.

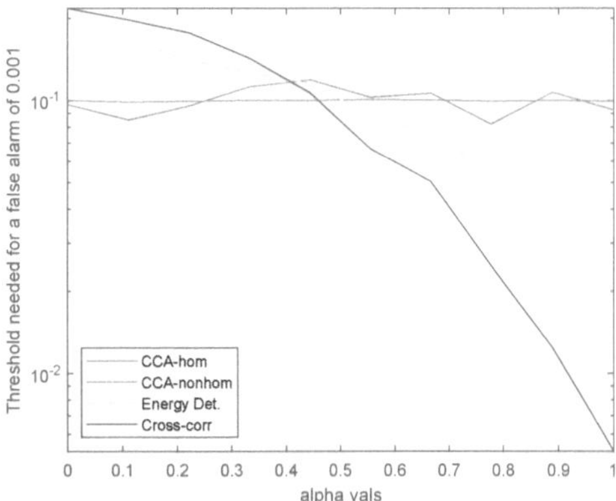

Figure 5.5: Required threshold to maintain $P_{fa} = 10^{-3}$ for the CCA-based and other detectors across various values of α.

Appendix A: A Lemma for Linear Signal Models

In this appendix, we establish a lemma used to achieve the maximum likelihood optimization required for the one-step GLRT derivation in Chapter 3.

Consider the linear signal model $\mathbf{X} = \mathbf{A}\boldsymbol{\Theta} + \mathbf{N}$ and define $\mathbf{Y} = \mathbf{X} - \mathbf{A}\boldsymbol{\Theta}$. Then, we claim

$$
\begin{aligned}
\hat{\boldsymbol{\Theta}} &= \arg\min_{\boldsymbol{\Theta}} \ \det\left(\mathbf{I} + \mathbf{Y}^H\mathbf{Y}\right) && \text{(A.1)} \\
&= \arg\min_{\boldsymbol{\Theta}} \ \operatorname{tr}\left(\mathbf{Y}^H\mathbf{Y}\right) && \text{(A.2)} \\
&= \arg\min_{\boldsymbol{\Theta}} \ \|\mathbf{X} - \mathbf{A}\boldsymbol{\Theta}\|_F^2 && \text{(A.3)} \\
&= \mathbf{A}^+\mathbf{X} && \text{(A.4)}
\end{aligned}
$$

where $^+$ denotes the Moore-Penrose pseudo-inverse. To show the claim, start by completing the square: let $\hat{\boldsymbol{\Theta}} = \boldsymbol{\Theta} - \mathbf{U}$. Then,

$$
\begin{aligned}
J_1(\mathbf{U}) &= \det\left(\mathbf{I} + \mathbf{Y}^H\mathbf{Y}\right) && \text{(A.5)} \\
&= \det\left(\mathbf{I} + (\mathbf{X} - \mathbf{A}\boldsymbol{\Theta})^H(\mathbf{X} - \mathbf{A}\boldsymbol{\Theta})\right) && \text{(A.6)} \\
&= \det\left(\mathbf{I} + (\mathbf{X} - \mathbf{A}\hat{\boldsymbol{\Theta}} + \mathbf{A}\mathbf{U})^H(\mathbf{X} - \mathbf{A}\hat{\boldsymbol{\Theta}} + \mathbf{A}\mathbf{U})\right) && \text{(A.7)} \\
&= \det\Bigg(\underbrace{\mathbf{I} + (\mathbf{X} - \mathbf{A}\hat{\boldsymbol{\Theta}})^H(\mathbf{X} - \mathbf{A}\hat{\boldsymbol{\Theta}})}_{\mathbf{M}_1} + \mathbf{U}^H\mathbf{A}^H\mathbf{A}\mathbf{U} && \text{(A.8)} \\
&\qquad\qquad + \underbrace{\mathbf{U}^H\mathbf{A}^H(\mathbf{X} - \mathbf{A}\hat{\boldsymbol{\Theta}}) + (\mathbf{X} - \mathbf{A}\hat{\boldsymbol{\Theta}})^H\mathbf{A}\mathbf{U}}_{\mathbf{M}_2}\Bigg) && \text{(A.9)}
\end{aligned}
$$

Note $\mathbf{A}^H(\mathbf{X} - \mathbf{A}\hat{\boldsymbol{\Theta}}) = 0$ by the orthogonality of the least-squares error, so $\mathbf{M}_2 = 0$. So we have,

$$J_1(\mathbf{U}) = \det\left(\mathbf{M}_1 + \mathbf{U}^H\mathbf{A}^H\mathbf{A}\mathbf{U}\right) \tag{A:10}$$

where both $\mathbf{M}_1$ and $\mathbf{U}^H\mathbf{A}^H\mathbf{A}\mathbf{U}$ are positive semi-definite, and $\mathbf{M}_1$ is not a function of $\mathbf{U}$. Thus, J_1 is minimized by $\mathbf{U} = 0$, and the claim is shown.

The result is a familiar one, and similar arguments are found, for example, in [58, 44].

www.ingramcontent.com/pod-product-compliance
Lightning Source LLC
LaVergne TN
LVHW041721190726
843493LV00007B/2191